Climate Change, Education, and Technology

Editors

Menşure Alkış Küçükaydın
Eregli Faculty of Education
Necmettin Erbakan University, Konya, Turkey

Hakan Ulum
Eregli Faculty of Education
Necmettin Erbakan University, Konya, Turkey

Ömer Gökhan Ulum
Faculty of Education
Mersin University, Mersin, Turkey
English Language Teaching Education

CRC Press is an imprint of the
Taylor & Francis Group, an **informa** business
A SCIENCE PUBLISHERS BOOK

First edition published 2025
by CRC Press
2385 NW Executive Center Drive, Suite 320, Boca Raton FL 33431

and by CRC Press
4 Park Square, Milton Park, Abingdon, Oxon, OXl 4 4RN

CRC Press is an imprint of Taylor & Francis Group, LLC

Library of Congress Cataloging-in-Publication Data (applied for)

ISBN: 978-1-032-69862-5 (hbk)
ISBN: 978-1-032-71965-8 (pbk)
ISBN: 978-1-032-71967-2 (ebk)

DOI: 10.1201/9781032719672

Typeset in Times New Roman
by Prime Publishing Services

Preface

The climate change crisis is undoubtedly one of the greatest challenges humanity has ever confronted. If we are to leave a habitable planet for future generations, we must support increasing local and global efforts to combat the crises. There is, in-fact, global movements attempting to find, explore and implement solutions. Given that human activities are one of the most significant causes of climate change, clearly, solutions must also come from humanity. Many sectors within society, including NGOs, universities, and municipal governments, are attempting to find such solutions. Educational institutions, especially, have significant responsibilities. Education is a requisite to increase awareness and, generate and implement solutions. Educating young people about the climate crises will substantially contribute to long term solutions. Technology too can contribute to this process by providing solutions, accumulating and analyzing data, and increasing energy efficiency. Furthermore, it is evident that young people are adept at using technology effectively. This book seeks to explore the intersection of and integrated approaches to climate change, education and technology. The book is an essential step in raising awareness, discussing solutions, and uniting those working in this field. In the context of Climate Change, Education, and Technology, this seeks to offer a comprehensive variety of perspectives, allowing for the generation of novel and inventive ideas. The collaboration of authors from various disciplines addressing the issues in question permits the emergence of novel and intriguing perspectives.

The target audience of the book is quite broad. It is a suitable resource for anyone interested in climate change, education and technology.

The primary audience are Universities, NGOs, Foundations and Associations, Climate Workers, Politicians, and Educators.

Furthermore, it is useful for academics, students, professionals in related fields, and general readers.

We would like to thank the authors of this book for their contributions and also the reviewers for improving the quality, coherence, and content presentation in all the chapters.

Editors

Contents

Chapter 1

An Interview with Colin Summerhayes

Climate Change

Manuel Varela, Ann Varela* and *Michael F. Shaughnessy*

Is there Any Credence to the Claim that Climate Change is not Caused by Human Activity?

Most people are aware that melting snow and ice adds water into the ocean, raising the sea level slightly. Fewer people know that the sea level also rises because water expands when heated. Fewer people still, realize that in the process we are losing Earth's refrigerator (it's like leaving your house to go on holiday and forgetting to close the freezer door—when you return, your food is rotten).

Indeed, very few people are aware that 90% of anthropogenic global warming is not in the air. It's in the ocean. The ocean has what physicists call a high heat capacity—It can absorb much more heat than the air. The top 3m of the ocean surface contains as much heat as the entire atmosphere. The heat absorbed by the ocean is slowly making its way to greater depths—down to 2000 m deep and beyond—causing even deep water to expand and further raise sea levels. As the rate of increase in

Eastern New Mexico University, Portales, New Mexico.
Emails: Manuel.Varela@enmu.edu; Ann.Varela@enmu.edu
* Corresponding author: Michael.Shaughnessy@enmu.edu

global average temperature has risen since 1900, so has the rate of rise in sea levels. It's now just above 4mm per year. There's a natural delay, a lag, if you will, between increasing greenhouse gases, rising temperatures, and rising sea levels. Eventually, on a timescale of 200 years or more, long after we stop adding CO_2 to the atmosphere, these three elements will reach an equilibrium. By then, global average temperature may have increased 3–4°C above the 1900 level, corresponding to an 8–10°C increase in the Arctic, and, consequently, a 10–15 m rise in global sea level.

Yet another significant effect of anthropogenic emissions is that CO_2 added to the air immediately mixes with the ocean surface to keep the air and water in chemical balance. Adding CO_2 to the ocean slightly reduces its alkalinity (or, in other words, makes it slightly more acidic). This effect is called ocean acidification. It poses a risk to forms of sea life that create their shells from calcium carbonate—among these are the pteropods or *sea butterflies* at the base of the food chain. There are significant dangers to our food supply in messing with ocean chemistry.

What Steps Can Local, State, and Federal Governments Take Toward Effective Legislation? What Can We, as Citizens, do About it?

So, what can be done about all this? Under the UN's Framework Convention on Climate Change (UNFCCC), governments have agreed that we should aim to be taking out of the atmosphere as much CO_2 as we are putting into it by the year 2050. The concept is termed *Net Zero*. It sounds wonderful, but it carries a significant downside. It means that by 2050, when atmospheric CO_2 will be substantially higher than it is now, it will then be stabilized at that high level. This, in turn, means that the climate will be much warmer than it is now, far more ice and snow will be melting, and the rate of rise in sea levels will be substantially higher than it is now. *Net Zero* is, in effect, nothing more than sticking a band-aid on a life-threatening stab wound (Summerhayes, 2024).

Climate scientists understand that there is currently less energy going back into space than there is coming in; the climate system is out of balance. What is needed is more heat energy dissipating from the Earth's atmosphere into space. This is only possible through a radical reduction in anthropogenic emissions of greenhouse gases. The term for such a reduction is *negative emissions*—bringing CO_2 levels down to, say, 350 ppm, which is where they were in 1988. The current level is

420 ppm and growing. Indeed, if we convert methane and nitrous oxide to their CO_2 equivalent, we currently have 500 ppm's worth of CO_2 in the air. Geology tells us that the last time this happened was 3.5 million years ago, in mid-Pliocene times (Summerhayes, 2020).

Fossil Fuels—What Role do they Play in Climate Change?

Evidently, we are facing an existential problem caused by the burning of fossil fuels (and more than 90 % of the fossil fuels ever burnt have been burned after 1950, during the lifetime of many people alive today). The only way to escape this is through new technologies that provide us with solar power, wind energy, nuclear energy, and hydrogen power (Ritchie, 2023). By ramping up onshore and offshore wind sources, the United Kingdom managed to create 50% of its electricity from wind and solar in 2022. More is promised! Stopping the sale of cars and trucks running on gasoline or diesel and encouraging people to buy hybrid or electric vehicles, is a move in the right direction. We require the diversion of large subsidies given to oil and gas companies into renewable energy sources, nuclear power and hydrogen. Stopping the purchase of new gas boilers in houses and replacing them with heat pumps is further a move in the right direction, as is converting old housing stock to become energy efficient and ensuring that all new housing is energy efficient from the word go. Battery technology is being improved as we speak. Wind and solar costs are now competitive with (or better than) fossil fuels. And, the latest research into hydrogen shows that it can be sourced remarkably cheaply by processing old plastics. Already, there are hydrogen-powered trains and hydrogen-powered construction vehicles (diggers and dozers).

What Definitive Scientific Evidence is there that Supports the Idea of Climate Change being a Valid Occurrence?

The climate changes for a number of reasons:

(*i*) the Sun's output of energy has increased by 30% over the past 4.5 billion years,

(*ii*) the shape of the Earth's orbit around the Sun changes in cycles of 100,000 and 400,000 years; when the orbit is more circular, our climate is warmer, but when it is more elliptical there are periods when the Earth is far from the Sun and our climate is cooler. These changes help to account for the recurrent ice ages of the past 2.6 million years during Pleistocene times. In turn, such changes are modulated by the precession of the Earth around its orbit, which leads to smaller cycles of heating and cooling roughly 20,000 years long and changes in the tilt of the Earth's axis 40,000 years long, changing the amount of heat reaching the poles (more tilt means more heat, which melts more ice) (Summerhayes, 2020). These cycles were studied by Milutin Milankovitch between 1920 and 1950 and are called *Milankovitch Cycles (Kerr, 1987)*. Currently, the influence of these cycles is more or less flat and likely to stay so for the next 5,000 years.

(*iii*) The Sun's output also varies cyclically; the most prominent cycles being those at 88 years, 208 years, and 2300 years. These have a much smaller effect than the long, slow Milankovitch Cycles. Since 1990, solar cycles (i.e., solar output in the short climate cycles) have been in decline.

(*iv*) Greenhouse gases absorb and reradiate heat (infrared radiation) emitted by the warmed surface of the Earth. They emit in all directions, thus warming the atmosphere. Geology tells us that there was running water on the surface of the early Earth around 3.8 billion years ago, so, the planet must have been warm enough for that to have occurred (Summerhayes, 2020). But, according to the laws of physics, the planet should have been extremely cold at the time. Only the existence of abundant greenhouse gases can account for this unexpected warmth. Hence, the so-called *Greenhouse Effect* is entirely natural. What humans have done with our emissions, is exacerbate the greenhouse effect, warming the planet more than it would otherwise have been by adding more greenhouse gases to the air. Experiments in 1856 by Eunice Foote in the USA (Foote, 1856) and in 1859 through 1870 by John Tyndall in the UK (e.g., Tyndall 1861, 1868; Archer and Pierrehumbert 2011), demonstrated that CO_2 is a greenhouse gas (as are methane and water vapor).

(*v*) Tectonic forces move the continents over time. When in polar positions, continental climate will be cool. When in tropical positions, it will be hot. However, these processes are extremely slow and not relevant to modern climate change.

(*vi*) Geologically, tectonic changes are sometimes associated with extreme volcanic activity and the resulting emission of unusually large volumes of CO_2. At such times, the geological evidence shows that the climate has been much warmer than in periods when such forces have been comparably weak. Eventually, this warming has led to acid rain that weathered mountain ranges, exposing minerals which absorbed CO_2 from the air and transferred it into the oceans, where it became trapped in limestones. Hence, we can see how changing CO_2 levels correspond with very slow, alternating periods of warming and cooling throughout the geological past.

(*vii*) Geology also tells us that the evolution, growth, and rapid spread of plants during early Devonian times, some 400 million years ago, removed CO_2 from the air. Also, the collection and decay of these plants in swamps created organic carbon traps. These two processes led to a cooling effect sufficient to bring about the Carboniferous-Permian Ice Age, some 300 million years ago.

(*viii*) Understanding the effects of these changes allows us to numerically model their combined effect on the climate.

What Biological Outcomes can be Expected for Specific Species of Living Organisms, for Example Microbes? What About Humans?

Current climate warming is affecting the distribution of plants, animals, and reptiles. In Europe, the growing season has become substantially longer than it was before 1950. Snowmelt in mountains ranges is causing Alpine plants to grow at higher altitudes. Marine organisms, plankton, and fish are now moving towards the poles. Coral reefs are increasingly subject to bleaching during marine heat waves. Humans are also experiencing more frequent and stronger heatwaves, which, in recent decades, have led to deaths in the tens of thousands in Western Europe and Russia during the summer months. Overheating has led to an increase in the number of wildfires, causing deforestation and the destruction of human settlements. Global warming also encourages insects to breed more frequently, which has led to effects such as the pine-bark beetle destroying much of Canada's pine forests. Continued warming is also likely to disrupt the monsoons, which will affect farmers across South Asia.

How is Climate Change Related to Evolution?

Any environmental change will lead to slow adaptation in organisms, with those most able to adapt being more successful than those that are less able. This is bound to occur with both warming and escalating ocean acidification. A major period of warming between the Paleocene and Eocene periods 56 million years ago, led to substantial changes in the nature of land animals and plants. Large quantities of CO_2 in the ocean, acidified bottom waters causing the demise of many bottom-dwelling creatures, among them benthic foraminifera. Many plants use what is called a C3 pathway, but grasses use a C4 pathway, allowing them to access CO_2 for growth more readily. The evolution of C4 grasses began during the Cenozoic period (which covered the past 50 million years), following the Cretaceous warm period (during which CO_2 was abundant due to tectonic activity—notably the breaking up of the Pangaea supercontinent and the dispersal of its fragments to their current positions), as the climate began to cool (Summerhayes, 2020).

Is There any Credence to the Claim Made by Climate Change Deniers that it is Not Human-Caused?

No. Their main argument is that the changes we are seeing are part of a natural cycle. However, geologists have abundant evidence as to how natural cycles have behaved over the past 11,700 years; what they call Holocene time. Over this period, there hasn't been anything like the abrupt global warming of the past 100 years (Summerhayes, 2024). The accelerated warming since 1970 (referred to above) is inextricable from the so-called *Great Acceleration* in multiple areas of human activity (Steffen et al. 2015). The critical factor has been tremendous growth in the human population between 1950 (when it was 2.5 billion people) to the present (8 billion people) and, the associated and equally tremendous surge in the use of fossil fuel energy to meet exponentially greater needs. The abrupt transition of the Great Acceleration from 1950 onwards has led us into a new geological epoch—the Anthropocene (Waters et al., 2016).

Simple physics and geological history make it clear that an increased amount of greenhouse gases warms the planet. The last time there was a major surge in CO_2 was about 56 million years ago, between the Paleocene and Eocene periods (mentioned earlier). Then, extreme volcanic activity produced enough CO_2 to cause the *Paleocene-Eocene Thermal Maximum*, which raised global average temperature by some 5°C and the sea level by approximately 10 m and altered ocean chemistry through ocean acidification.

How can Scientists Best Deal with Deniers?

It seems to me that one of the main questions we all face is one of values (Carney, 2021). What sort of life do we want for our children, grandchildren, even great-grandchildren, and beyond? What sort of world do we want to bequeath to them? Do we want to leave behind a world in which the tropics are too warm to work in, or one where the sea level around our coasts has risen by 10-15m on average (as geologists can confirm has happened in the past when conditions were extremely warm)? Do we want them to experience a major expansion of arid and desert areas (including all Mediterranean countries) and more frequent, larger wildfires every year? Do we want those who live in humid temperate regions to be frequently afflicted by tropical downpours that cause major flooding? Would it not be better to simply work hard at changing our energy supplies so that such dire consequences simply do not come to pass?

It occurred to me that it might be useful if I explained what the denialist community is saying about anthropogenic global warming. These are not only talking points, but common misconceptions. A primary argument, commonly from geologists, is that *the climate is always changing*. They tend to focus on the changes that occurred over the Pleistocene Ice Age of the past 2.6 million years, and, particularly, on data extrapolated from long Antarctic ice cores that drill back up to 800,000 years of climatic history. Some of this data indicates that CO_2 followed, rather than preceded, temperature change, which means, to these people, that CO_2 could not have been a driver of climate change, but rather that temperature was driving CO_2. This perception is based on the notion that, in a way, CO_2 and temperature are tightly twinned. However, we know from the longer geological record as well as from laboratory experiments that there can be times when temperature precedes CO_2 or when CO_2 precedes temperature. Let's focus on the ice core data. During the Pleistocene, CO_2 was at its lowest level for millions of years, ranging from 280 ppm during the warmest interglacial period down to just 180 ppm during recurrent glacial maxima. Given that geological history shows that this was a period of unusually low CO_2, temperature was the driver of change due to Milankovitch Cycles caused by changes in the Earth's orbit and the tilt of its axis (as explained earlier). When temperature cools, several environmental changes occur: (*i*) atmospheric CO_2 is absorbed by cooling oceans (warm water releases dissolved gases, while cool water absorbs them), (*ii*) ice sheets form, causing sea levels to fall, so there is less water surface available to absorb CO_2, (*iii*) sea-ice covers the polar oceans, further reducing water surface area and thus the quantity of CO_2 absorbed. Hence, CO_2 levels do not mimic the declining temperature or do so weakly. Over several tens of thousands of years, CO_2 levels eventually *catch up* with the flattening profile of temperature (Summerhayes, 2020, 2023). The geologists who deny that CO_2 plays a key role in climate change take these processes, wrongly, as evidence that CO_2 does not drive temperature.

The latest data (2013 onwards) shows that as Milankovitch cooling phases subside and their warming phases begin, CO_2 and temperature rise in tandem. This is because oceans warm, releasing CO_2, land-ice melts, raising sea levels, and sea-ice reduces in area—melting sea-ice increases the water surface area available for CO_2 to move from the ocean to the air. The net results of these interactions can be seen in the ice core data, marine and lake sediment cores, and stalagmites and stalactites in caves. We understand these interactions and the results they bring about. The data does NOT support the argument that temperature *always* drives CO_2.

Geologists making such arguments choose to ignore evidence from the Paleocene-Eocene transition, 56 million years ago, when several massive eruptions caused by the opening of the northern North Atlantic, expelled large quantities of CO_2 into the atmosphere, leading to similar results to those we are seeing today —a rise in CO_2 driving a rise in temperature. The rates at which CO_2 and sea levels rose then were far slower than they are today, taking place over several thousands of years, but the process was analogous. The so-called Paleocene-Eocene Thermal Maximum that resulted is clear evidence that CO_2 can drive climate.

The same geologists also tend to ignore the long record of temperature and CO_2. Using a variety of paleoclimatic indicators, we can track temperature back at least 500 million years and CO_2 levels back to about 450 million years, as documented in Summerhayes (2020, 2023). This data demonstrates the strong link between CO_2 and temperature over vast lengths of geological time. An instructive example is the decline in temperature that's taken place from the peak warmth of Eocene time, 50 million years ago, to the present. The cause being the gradual post-Cretaceous decline in seafloor expansion, driven by plate tectonics, and the resulting gradual rise of mountain ranges. The decline in seafloor expansion and associated decline in CO_2 production resulted in a gradual cooling. The development of mountains facilitated the weathering of minerals by the carbonic acid in rainwater, leading to the absorbtion of CO_2 from the air. The decline in CO_2 resulted in the formation of the first Antarctic ice sheet 34 million years ago (no external driver caused the cooling—only CO_2 and water vapor). The world continued to cool, leading, some 2.6 million years ago, to the development of the Northern Hemisphere ice sheets that once covered Scandinavia, Iceland, Greenland, northwest Europe (including Britain), and the northern half of North America.

Climate scientists understand that doubling the amount of CO_2 in the atmosphere will inevitably lead to a substantial increase in temperature. This is not an article of faith. It is based on our understanding of the results of experiments conducted in the mid-1800s (clearly showing that CO_2, methane and water vapor are greenhouse gases and that they both absorb and re-emit infrared radiation). These experimental results are confirmed by observations of the climates of other planets (Mars and Venus), by observations (i.e., measurements) of what is happening in our own atmosphere, and by numerical calculations of the effects of changing the amounts of CO_2 (and other greenhouse gases) in the atmosphere. Some physicists (not climate scientists) argue that doubling the amount of CO_2 in the air will not cause warming. However, for the most part, their expertise lie outside of climate science. They appear to be unaware of the extensive and seminal work done on planetary climate change by the eminent Oxford University physics professor Raymond Pierrehumbert (Pierrehumbert, 2010, 2011) Some denialists like to argue that a mere 0.04 percent of the atmosphere (the prevalance of CO_2 in the air) cannot possibly influence our climate. The same percentage of alcohol in one's blood would have them arrested for drunken driving. Small quantities can indeed have large effects. Clearly, their arguments ignore Professor Pierrehumbert's explanations of planetary climates.

Denialists also argue that the carbon-14 (^{14}C) added to the air by the interaction of cosmic rays and nitrogen gas molecules, disappears in about fifteen days, the implication being that any CO_2 we put into the air will do the same. But, this is mixing apples and oranges. While it is true that an individual CO_2 molecule introduced into the atmosphere has a short residence time , the same molecule introduced into the ocean will result in a reciprocal CO_2 molecule being released into the air from the ocean (in this case, the small vanishing amounts of the ^{14}C CO_2 molecules would be replaced in the air by reciprocal counterparts from the very large oceanic reserve of ^{12}C CO_2 and the smaller reserve of ^{13}C CO_2). There is a constant exchange of CO_2 across the air-sea interface to ensure that the two environments maintain a chemical equilibrium (

with regards to CO_2). It is a fact that the quantity of CO_2 in the air is increasing gradually, year-on-year, due to our emissions. Approximately twenty-five percent of those emissions end up in the ocean, so oceanic CO_2 is also gradually increasing. The rate of increase would be greater if CO_2 from the air dissolved rapidly down to greater depths, but this is an extremely slow process leading to the ocean gradually *filling up* with CO_2 The CO_2 we have put into the air will be with us, in the air and oceans, for hundreds of years.

We may ask ourselves, "what is it that drives deniers?" Are they influenced by and do they benefit from fossil fuel industries? Do they feel we are interfering with God's plan for the planet? Do they ascribe to a biblical injunction to exploit the Earth instead of one to act as stewards of the Earth? Are they extreme individualists who believe that we should be allowed to do whatever we like to the planet regardless of the consequences?

All scientists are skeptical by training. Even after applying that trained skepticism, climate scientists from our major scientific institutions tell us that there is something to be concerned about. Nevertheless, it is the case that we are living in an age characterized in some quarters by an assault on reason (Gore, 2007), and by mistrust of science (Oreskes, 2019). Despite that mistrust, there are abundant reliable sources of the necessary information. However, much of this invaluable information is invisible to the public and the media because the results of most modern research work on the climate system by a multitude of climate scientists in academic and governmental institutions are largely found in scientific journals that are located behind the publishers' pay-walls, making the data inaccessible to the public. This is why I write books, to bring that hidden knowledge out into the open (Summerhayes 2020, 2023).

References

Archer, D. and Pierrehumbert, R. (2011). The *warming papers:The scientific foundation for the climate change forecast.* Wiley-Blackwell.

Carney, M. (2021). *Values: Building a better world for all.* William Collins.

Foote E. (1856). Circumstances affecting the heat of the sun's rays. *American Journal of Science and Arts, 22*, 382–83.

Gore, A. (2007). *The assault on reason.* Bloomsbury.

Oreskes, N. (2019). *Why trust science*? Princeton University Press.

Pierrehumbert, R.T. (2010). *Principles of planetary climate.* Cambridge University Press.

Pierrehumbert, R.T. (2011). Infrared radiation and planetary temperature. *Physics Today, 64*(1), 33–38.

Ritchie, H. (2024). Not the end of the world. Penguin, Random House, 340 pp.

Steffen, W., Broadgate, W., Deutsch, L., Gaffney, O. and Ludwig, C. (2015). The trajectory of the Anthropocene: the great acceleration. *The Anthropocene Review*, *2*(1), 81–98.

Summerhayes, C. P. (2020). *Paleoclimatology: From Snowball Earth to the Anthropocene*. John Wiley & Sons.

Summerhayes, C. (2023). *The Icy Planet: Saving Earth's Refrigerator*. Oxford University Press.

Summerhayes, C.P. (2024). Net zero is not enough. *Geoscientist, 34(*1), 18–23.

Tyndall, J. (1861). On the absorption and radiation of heat by gases and vapours, and on the physical connexion of radiation, absorption, and conduction. *Philosophical Magazine and Journal of Science, 22*, 169–194. https://doi.org/10.1098/rstl.1861.0001

Waters, C.N., Zalasiewicz, J., Summerhayes, C., Barnosky, A.D., Poirier, C., Gałuszka, A. and Wolfe, A.P. (2016). The anthropocene is functionally and stratigraphically distinct from the Holocene. *Science*, *351*(6269), aad2622.

CHAPTER 2

Educational Escape Rooms in Climate Change Education

Learning by Doing

*Tania Ouariachi** and *Wim J.L. Elving*

Introduction

The fight against climate change is a necessity for a sustainable future. Transitioning between energy sources is a matter of utmost importantance. Creating clean, affordable energy is one of the main challenges the world faces today; how can we provide the energy needed to support growing populations and economies, without damaging the environment beyond repair? The need for an energy transition has become increasingly dire in context of centralized energy systems highly dependent on fossil fuels, oil and gas prices soaring to new highs (crises such as that in Ukraine bringing new levels of uncertainty), and widespread unaffordability of energy bills and consequent energy poverty.

Raising awareness and educating young people on the complex issues of climate change, sustainability and energy transition, is an often-overlooked factor in the fight against climate change; topics perhaps challenging for students from diverse academic backgrounds. Young

Centre of Expertise Energy, Hanze University of Applied Sciences, the Netherlands. Email: w.j.l.elving@pl.hanze.nl

* Corresponding author: t.ouariachi.peralta@pl.hanze.nl

people have tremendous potential to play an active role in the fight against climate change however, in institutions of higher education, the question remains on how best to educate, motivate, and empower youth to become agents of change and inspire the same in others.

Educational escape rooms have become a popular way to encourage experiential learning; engaging people through a learning environment. Escape rooms can be defined as "a live-action team-based game where players discover clues, solve puzzles, and solve tasks in one or more rooms to accomplish a specific goal (usually escaping from the room) in a limited amount of time" (Nicholson, 2015). According to Ball (2012), the use of gamification approaches in higher education institutions, could be a remediating response to an environment of increasing "performativity" and "instrumentalism".

While many initiatives are undertaken in universities, they are often led by teachers and focus on sustainability in only a broad, general sense (Ouariachi and Elving, 2020). It is in this context that the project *Beat the clock, turn the lock* was born at Hanze University of Applied Sciences, where students from the Master Energy for Society program (20 students from different backgrounds and nationalities) design an educational escape room. Through this game, they were expected to reinforceinformation on energy issues, apply theoretical course content on communication and behaviour, and put into practice important skills such as creativity, problem-solving, collaboration, and critical thinking. The result is an escape room experience, open for other students from the University to play and subsequently learn themselves about energy transition in a fun way. Using the pedagogical concept of *Learning by Design*, students, with the support of their instructors, achieve engage with concepts through cycles of designing and testing, rather than the conventional memorizing of facts to be reproduced on tests.This paper presents the set-up and implementation of the project and the resulting research of qualitative surveys assessing the impact on student learning. We end with observations from the experience and lessons learnt to inspire other academics and eductors interested in such novel and innovative pedagogy.

Connecting Climate Change Education and the Energy Transition

Climate change has devastating effects on everyday life. In recent years, we have experienced increased occurrences of flooding, wildfires, extreme heatwaves, and other extreme weather conditions. We use the *A to Sustainability* model (see Figure 1) to inspire individuals to start taking (personal) action regarding climate change.

The model begins by instilling a sense of urgency. Individuals lacking this sense of urgency should be presented with relevant information and facts on climate change, for instance Intergovernmental Panel on Climate Change (IPPC) reports. The IPCC prepares comprehensive Assessment Reports about the state of scientific, technical and socio-economic knowledge on climate change, its impacts and future risks, and options for reducing the rate at which climate change is taking place. If, as many by now do, they realize this urgency, the following step is creating awareness. Awareness is the most important phase of the model. It entails cognisance of personal behaviours and habits, for example energy consumption of home appliances. Few people realize how much energy their refridgerator consumes, and when they happen to know the quantity, they often don't know what it means (exactly how much is 300 KwH per year?). In various research projects implemented in the context of the energy transition in The Netherlands, like *Check je Energieslurper (Check your energy guzzler)* and *Bedrijfkracht (Company Power)*, teams were formed (e.g., neighbours, friends, colleagues) that used energy meters, and team members checked and compared the data from the energy meters with each other: the data gathered from the questionnaires and the debriefings after the sessions revealed that some people in those teams immediately bought themselves a new fridge after learning that their fridge was using as much as three times more energy compared to that of other team members (Wikens, 2019).

Therefore, raising awareness is followed by taking action, individual or collective, and subsequently by dialogue, which is a symbolic phase needed for a sustainable future. Local Energy Initiatives (LEI) are

Figure 1: From A to Sustainability (Elving, 2020).

examples of collective action and dialogue. They are described by Oteman et al. (2014: 2) as: "decentralized, non-governmental initiatives of local communities and citizens to promote the production and consumption of renewable energy". In a LEI case study described by Soares da Silva and Horlings (2020), the dialogue established with residents in nearby villages helped to create new relations within the area, especially between city dwellers (favourable to the project) and villagers (who were concerned with the project). One of the initiators mentions that through the community engagement, people who live in the area became interested in energy and sustainability issues, even though they were first against the project. The purpose of educational escape room games is to raise awareness among audiences that may otherwise be hard to reach, such as people from non-academic backgrounds and young people, who seldomly take part in such energy projects (Mees, 2022). Games and escape rooms allows us to reach these demographics; being invited to play a game, rather than being force-fed information, is often a more attractive prospect.

Learning by Design

The *Learning by Design* (LBD) approachis derived from *Constructivism* learning theory. Central to the constructivist approach is the idea that learners construct knowledge by building on what they already know. Research reveals that students learn best from constructivist methods; concrete experiences relaying information that serves as a basis for reflection and is subsequently better assimilated facilitating the formation of abstract concepts (Taber, 2011). The activity of designing an educational escape room implements the *LBD* approach by allowing students to learn experientially, through design challenges, rather than the conventional memorization of facts, formulas and information to be later reproduced on tests. Students become immersed in the concepts being covered and learning occursthrough a cycle of designing, testing, redesigning and explaining.By working iteratively to improve upon their design solutions, they enhance their understanding of concepts and get a chance to put into practice a variety of skills. Throughout, the design challenge provides the glue that connects inquiry, investigation, drawing conclusions, and application (Kolodner et al., 2023).

There has been little research on the application of the LBD approach through gamification, with students beingthe game designers. More is needed. While escape room design has been implemented prior in creative courses where students become "makers" (Dougherty, 2012) and support to develop the relevant underlying creative skills (Li et al., 2018), here, the creation was a means to an end: designing an escape room to master the content themselves. When talking about energy transition, initiatives are almost absent, and most of these energy transition-related escape rooms are still in their early stages and in the form of traditional puzzles. In addition, just few of them follow a bottom-up approach and are designed by students. Therefore, we know little about the effects of these types of strategies to engage students in the energy transition. This is especially relevant because youth has a huge potential to play an active role in the energy transition.

Beat the Clock, Turn the Lock

The project *Beat the clock, turn the lock* (with the support of the Comenius Teaching Fellowship—Dutch Government and the Centre of Expertise Energy—Hanze University of Applied Sciences) was implemented in the second phase of the Communication & Behaviour course within the Energy for Society master's program. Twenty students of different nationalities and with a range of academic backgrounds took part in this master's program. The first part of the course consisted of theory lessons on behavioural change, communication strategies, storytelling and persuasion. In the second phase, the class was divided into teams of four students tasked with the development of an educational escape room. Each team was required to follow a process of design, creation and testing over several weeks. The following graph (Figure 2) illustrates the design process:

Each team had to develop a different mission (story and puzzle) for their escape room. The missions were to be centred around a specific concept related to energy transition (and consequently a unique learning outcome for each topic). Some themes covered were consumption, production, storage and transformation. In the process, students applied theoretical knowledge from their classes and put into practice important skills such as creativity, problem-solving, collaboration, and critical thinking. Throughout, they received instruction and coaching from experts in game design, energy, psychology and communication.

Each team crafted their missions into a final prototype. The escape room game was then played and tested, first by the class, and then, at a held event, by a sample group of studentsrepresenting the main target group (the student body at Hanze University of Applied Sciences),

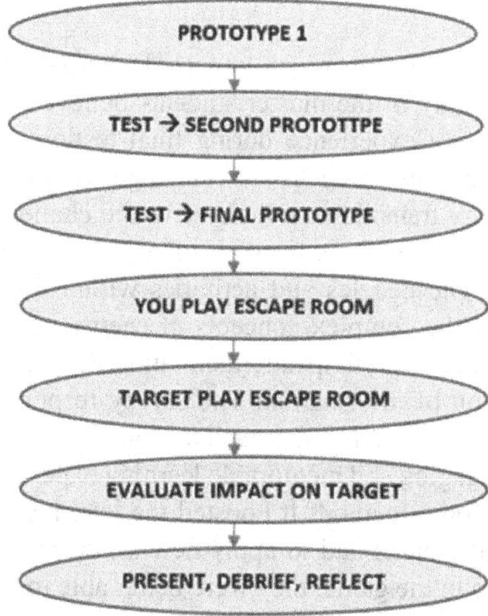

Figure 2: Design process (own elaboration).

where impact and outcomes could be analyzed. A final debrief between the coaches and students to reflect on the experience was conducted to conclude the exercise.

Methodology

The study conducted to analyse the effects of the educational escape room is qualitative in nature. Qualitative surveys, making use of open-ended questions, are used as a method of data collection to ascertain the impact of the project on student learning. The surveys ask the students to reflect on 1) what affect does the escape room experience have on players' learning about energy transition; 2) how does the escape room engage players on cognitive, emotional, and behavioural levels; 3) how has the escape room project impacted personal learning about the energy transition; 4) how can such projects help to create a sustainable society. The surveys were submitted at the conclusion of the course, after the final prototypes had been played and tested, and students had a chance to observe, analyze and reflect on the results. The sample of 20 students who participated in the project were aged between 21 and 30 years-old, had diverse academic backgrounds (e.g., engineering, law, psychology, etc.) and several nationalities (e.g., Dutch, Taiwanese, Mexican, etc.).

Results

When asked how the escape room impacted players' learning about energy transition, many of the master students believed that there was an observable positive experience during final testing. They felt that designing and playing the escape room challenge was an engaging way to learn about energy transition. In their view, the challenge of solving puzzles required team members to work together to decode clues and execute activities, The puzzles and activities within the escape room helped break down the complex concepts of energy transition making them more accessible to participants, facilitating deeper understanding and greater retention of information. The survey responses stated that by using a hands-on, interactive approach to education, the escape room provides a more engaging and memorable learning experience compared to traditional classroom methods.It boosted the learning experience by allowing Since participants had to apply new knowledge immediately, in order to progress in the game, they were better able to grasp complex concepts. It was also noted that the use of storytelling created a fun and engaging environment that increased player motivation and interest in the subject matter. Another interesting observation highlighted in in the survey responses was that the experience differed between those players who had previously experienced an escape room and those who had not. Also, some players, who were studying renewable technology, were better able to complete some of the challenges.

When asked how the escape room experience engaged players on a cognitive, emotional, and behavioural level, most students could appreciate the cognitive and emotional impact, but not so much the behavioural. Cognitive engagement was attributed to elements such as the challenging puzzles and brain teasers which required critical thinking and problem-solving skills. Clues and riddles requiring active reasoning, combined with time pressure which demands quick decision-making significantly enhanced cognitive engagement. Emotional engagement occurred through iwell-crafted storylines that created a sense of urgency and immersed participants in the experience. The feelings of accomplishment and pride when puzzles were solved and obstacles overcome, alongside immersive elements such as the use of dioramas, UV lighting, and feedback sound in the form of a buzzer also helped in achieving this aim.

When asked to reflect on how designing and playing the escape room experience impacted their own learning about the energy transition, here are some direct quotes from the game designers:

> *"By creating a puzzle for others, you have to really put yourself in the shoes of the player(s). How can we effectively transfer knowledge in an understandable way? I learned that it is essential that the energy transition and its problems first must be simplified for people"* (student 1).

> *"To start off, I think it was a good way for all the participants in this project to start by experiencing the escape room themselves. Because we played an escape room where everyone goes in blank with no benefits, you are directly placed in the position of our final target group. The escape room project made me aware of the importance of knowing what you want to bring to your target group. It is nice to have seen and experienced a setting where your target group also will be placed in"* (student 2).

> *"The fun and interactive nature of the escape room increased my motivation to learn about the topic and increased my awareness about the importance of reducing energy consumption. For example, by comparing the energy usage of different lamp types, I found out that LED ones are more efficient than other lamps, and also that vacuum cleaners use much more energy rather than other appliances*" (student 3).

> *"I found out that the escape room mission encourages collaboration and teamwork, promoting the idea that energy transition is a collective effort that requires the participation of everyone"* (student 4).

> *"Given my prior education, experience, and involvement in energy transition initiatives, I found it challenging to pinpoint specific new insights gained during this course about the energy transition itself. Instead, my focus was primarily on constructing a functional puzzle for the escape room. However, I now have a greater appreciation for the potential role of hydrogen in the energy transition, surpassing my previous understanding"* (student 5).

Finally, the game designers were asked to reflect on how designing a mission, using innovative methods and ideas, can help to create a sustainable society. Here are some representative answers:

> "*By using innovative ways of creating energy transition awareness, you can attract a larger audience. A lot of people play escape rooms for fun, not necessarily for the educational benefit. But, by incorporating this benefit, you reach people who maybe don't think about energy transition often*" (student 1).

> "*At first, I was sceptical when I heard about the plan to create an escape room experience for the participants to become more aware of the energy transition, and maybe even influence their behaviour. But after having finished the project and listening to the experiences of the first groups that played the escape room, I'm convinced that looking outside of the box for interventions can help create a sustainable society. I believe that creating an experience that's memorable, immersive, and fun can have a great impact on someone's perspective on the energy transition*" (student 2).

> "*Designing an educational escape room experience using innovative ways can contribute to creating a sustainable society in several ways. Firstly, the escape room format can help to raise awareness about energy consumption and the impact it has on the environment, encouraging players to think about their own energy usage and how they can reduce their carbon footprint. Secondly, by making players actively involved in learning about the energy transition, the escape room can encourage more sustainable practices in daily life. The escape room challenge fosters collaboration and teamwork, emphasizing the belief that energy transition must involve the joint efforts of all individuals*" (student 3).

Overall, the students appreciated the opportunity to participate in the escape room project. It was the experience was perceived as a fun, but also stressful and challenging, due to the time deadlines and the logistics of building a physical game (crafting, constructing, shopping, etc.).

Discussion and Conclusions

In institutions of higher education, there has been an increasing trend towards innovative pedagogy in recent years, but a focus on the energy transition issue has been lacking, making the escape room project original and perhaps unique. Crucially, the escape room experience was designed by students for students, following a Learning by Design approach, which is a rarity given that mostly teachers take the lead in activity development (Ouariachi and Elving, 2020). In the context of this project, the success factor seems to be the existence of a tangible challenge; creating an escape room that is going to be played by others. A rented shipping container was available to create the game, making the challenge tangible and more urgent. In addition, it was also noted that the use of storytelling created a fun and engaging environment that increased player motivation and interest in the subject matter, as corroborated by Voloshynov et al. (2020) and Kiryakova et al. (2014).

On the other hand, more time devoted to coaching and reflection sessions during the activity , less work related to logistics, and a better course structure, would have made the project more effective. It is important to consider that such a pedagogical approach is usually preferred compared to traditional classes, as corroborated by Manzano-León et al. (2021), however, it may not appeal to all students (as can be seen in the scepticism of some answers above), so it is important to adjust according to student needs and learning styles and find the correct balance. To enhance learning in LBD projects, Kolodner et al. (2003) offer insights in how the process can be structured, suggesting fostering interaction between staff and students and between student groups, through methods like class discussions, pin-up sessions and gallery walks. These stimulate feedback being given and the learning be made explicit. By listening to learners' experiences and incorporating their feedback, instructors can tailor instructional materials to meet their needs better and enhance engagement and learning outcomes.

Though this project took place in the realm of education institutions, it clearly demonstrates the opportunities inherent in gamification approaches to engage and inspire the public towards taking climate action. Gamification is a powerful tool to raise awareness. Projects adopting such an approach may be able to engage groups of people who, in other circumstances, may not be inclined to act, either becausethey are unaware of what actions they can take , or are simply indifferent towards the topic. In playing the escape room, players are engaged and gain knowledge on energy related issues, like energy saving and insulation. We can hope that this knowledge will then be applied in peoples' homes and communities. Of course, further tracking and study is needed over a long-term period.

For future research, we recommend experimentation on larger cohorts to refine and further establish conclusions, analyse differences between students from different cultural backgrounds, and engage in comparative studies between a *traditional* pedagogies and game-based learning.

References

Ball, S.J. (2012). Performativity, commodification and commitment: An I-Spy Guide to the Neoliberal University. *British Journal of Educational Studies, 60*(1), 17–28. https://doi.org/10.1080/00071005.2011.650940

Dougherty, D. (2012). The maker movement. *Innovations, 7*(3), 11–14.

Elving, W.J.L. (2020). Sustainable Communication. *International Journal Environmental Sciences & Natural Resources 26*(3), 90–92. https://doi.org/10.19080/IJESNR.2020.26.556190

Kolodner, J.L., Camp, P.J., Crismond, D., Fasse, B.B., Gray, J., Holbrook, J., Puntambekar, S. and Ryan, M. (2023). Problem-based learning meets case-based reasoning in the middle-school science classroom: Putting learning by design(tm) into practice. *The Journal of the Learning Sciences, 12*(4), 495–547. https://doi.org/10.1207/S15327809JLS1204_2

Li, P., Chou, Y., Chen, Y. and Chiu, R (2018). *Problem-based Learning (PBL) in Interactive Design: A Case Study of Escape the Room Puzzle Design* [Conference]. 1st IEEE International Conference on Knowledge Innovation and Invention (ICKII), Jeju Island, South Korea.

Manzano-Leon, A., Camacho-Lazarraga, P., Guerrero, M., Guerrero-Puerta, L., Aguilar-Parra, J.M., Trigueros, R. and Alias, A. (2012) Between Level Up and Game Over: A Systematic Literature Review of Gamification in Education. Sustainability, 13(4), 2247; https://doi.org/10.3390/su13042247

Mees, H.L.P. (2022). Why do citizens engage in climate action? A comprehensive framework of individual conditions and a proposed research approach. *Environmental Policy and Governance, 32*(3), 167 – 178. https://doi.org/10.1002/eet.1981

Nicholson, S. (2015, October). *Peeking behind the Locked Door: A Survey of Escape Room Facilities.* White Paper [online], http://scottnicholson.com/pubs/erfacwhite.pdf

Oteman, M., Wiering, M.A. and Helderman, J.-K (2014). The institutional space of community initiatives for low-carbon energy: a comparative case study of the Netherlands, Germany and Denmark. *Energy Sustain Soc, 4*(11). https://doi.org/10.1186/2192-0567-4-11

Ouariachi, T. and Elving, W.J.L (2020). Escape rooms as tools for climate change education: an exploration of initiatives. *Environmental Education Research*, 1–14. https://doi.org/10.1080/13504622.2020.1753659

Soares da Silva, D. and Horlings, L. (2020). The role of local energy initiatives in co-producing sustainable places. *Sustainability Science*, *15*, 363–377. https://doi.org/10.1007/s11625-019-00762-0

Taber, K. (2011). Constructivism as educational theory: Contingency in learning, and optimally guided instruction. Editor: J. Hassaskhah (Ed.), *Educational Theory*, 39–61.

Wiekens, C. (2019). Duurzaam Gedrag (Sustainable Behaviour). Inaugural Speech. Groningen: Hanze University of Applied Sciences.

CHAPTER 3

Current Technological Approaches to Climate Change Education

Applications and Implications for the Future

Menşure Alkış Küçükaydın

Introduction

Global climate change demands immediate attention. Societies and governments must develop adaptive policies and conduct proactive mitigation measures to overcome the threat (Jain et al., 2023). Education is necessary to raise awareness and support for such policies (Gautier and Solomon, 2005). In order to do so, we must view education and technology as being closely linked. In order to combat global climate change, first of all, information is needed. For example, where, how and how much greenhouse gases are emitted? Without this information and continuously monitoring greenhouse gas sources, that is, without recognizing the "enemy", fighting it is impossible. One of the most effective methods of recognizing the "enemy" at the moment is technology.Modern technology, 21st-century technologies such as *The*

Necmettin Erbakan University, Konya, Turkey.
Email: mensurealkis@hotmail.com, ORCID

Internet of Things, Earth observation satellites, cloud computing and artificial intelligence can primarily provide comprehensive information on greenhouse gas sources and quantities and near-instantaneous monitoring of greenhouse gas evolution. There are also technological solutions that are more narrowly targeted at local administrators but whose results can significantly contribute to combating global climate change (Cheng et al., 2017). Implementing such ideas can play a significant role in preventing environmental damage and promoting public awareness.

Contemporary researchindicates that technology-enabled educational methods better engage students (Swarat et al., 2012) and improve academic performance by deepening comprehension (Bodzin et al., 2013). The World Economic Forum (2022) reported that information communication technologies (ICTs) are transforming the globe and affecting sustainability. Technology can be utilised to combat climate change and improve global education. Teachers at all levels may increase climate change awareness by carefully designing curriculums and using relevant technologies.

The need for better climate change education has provided an opportunity to consider technological approaches. Researchers can benefit from knowing published and tried technology to improve findings. Previous climate change literature reviews have been attempted. Technology was not applied in these endeavours (Monroe et al., 2019; Wibeck, 2014). They concentrated only on climate change and its major themes (Hestness et al., 2014).

Technological Applications

Remote sensing technologies

Satellite remote sensing technologies enable the study of the Earth, its oceans, and atmosphere (Achieve, 2013). A study with university students used remote sensing technologies to map issues like deforestation in Brazil and global CO^2 levels. Pre- and post-course surveys reported that students were better able to engage with learning processes involving global data, and subsequently were better equipped to design research projects at appropriate scales (Cox et al., 2014).

Simulations

Simulations can be used in almost any discipline and level of education, including climate change. Reportedly, the use of simulations has resulted in improved students comprehension of climate research processes (Bush et al., 2018). Gautier and Solomon (2005) conducted a study with university students to examine the movement of solar radiation in the atmosphere and asked them to generate scientific questions using simulations. After a 10-week experimental period, it was observed that students were able to design specific experiments for radiative processes.

Artificial intelligence

Although remote sensing technologies are often preferred in climate studies, artificial intelligence (AI), which provides the opportunity to examine comprehensive spatial and temporal details of the Earth, has recently built a great deal of speed (Cox et al., 2014). AI, which collects satellite photographs of both distant planets and the Earth, has revolutionised image analysis (Dimitrovski et al., 2023a). AI is used to map land use and forests, monitor natural catastrophes, and chart precision agriculture (Cheng et al., 2017). Such data enables a wide range of weather, climate, and environmental studies. Dimitrovski et al. (2023b) completed the most extensive study; *AiTLAS*. Their Earth observation AI platform standardises data collection and processing. A comparable investigation, *EarthNets*, was done by Xiong et al. (2022). Through EarthNets, which uses artificial intelligence and remote sensing, over 400 land usage data sets, disaster monitoring capabilities and, climate change and weather forecasting models have been made available. Kulp and Strauss (2018) have examined the impact of sea level rise on coastal infrastructure using artificial intelligence simulations. Their application utilizes machine learning algorithms to make predictions on how roads, bridges, and buildings will be affected by rising sea levels. The algorithms predict whether certain roads or buildings will be at risk of flooding or whether certain areas are more susceptible to erosion due to rising sea levels.

Modeling technologies

Models are widely used in education to give students a more direct experience of scientific activity, to demonstrate how scientists work, to explore phenomena and related ideas, and to foster the ability to develop scientific argumentation (Campbell and Oh, 2015). Enabling students to grasp the underlying logic of the scientific endeavour is essential to an effective science education (Bell et al., 2010). In climate education, modeling technologies have become quite popular. The most detailed study in this arena was conducted by Bush et al. (2018). Bush et al. (2018) conducted experimental research over seven weeks using the Educational Global Climate Model modelling technology.A comparison group received climate change education using simpler tools. The study analysed learning trajectories of both groups of students; experimental and traditional. It was observed that the experimental group was at a significant advantage with higher levels of engagement achieved, stronger learning trajectories, and a better understanding of how climate scientists conduct their research.

Virtual reality technologies

Virtual reality technology creates a 3D world that the user experiences as almost real. In the *VR* space, users can interact with their environment using sensory organs (Dede et al., 2017). Virtual world applications can trigger intrinsic motivation by providing students with a meaningful learning context (Lee et al., 2010). Makransky and Peterson (2019) reported that virtual reality technologies enhanced climate change learning by facilitating *high-level* learning. Reportedly, virtual environment learning increases attention and improves critical thinking (Schott and Marshall, 2018). Ou et al. (2021) created a virtual wetland ecological system using *VR 360°* panoramic technology for a grade-7 environmental education course. They then conducted a study to assess the impact on students. Following the virtual lessons, it was observed that students' motivation to learn more about climate change had increased, leading to the conclusion that VR is a useful tool for environmental education.

Another implementation of virtual reality in environmental education was conducted by Ruan (2022). Undergraduate students were trained with the *EduVenture* VR application and their attitudes towards environmental issues were assessed. According to the results, student awareness of environmental protection and sustainable development increased through the application. Markowitz et al. (2018) conducted an experimental design using virtual reality with 19 students aged between 16 and 18. The study showed that students who underwent the VR experience increased their knowledge of ocean acidification, leading to higher environmental attitude scores compared to their peers who had not. The immersive virtual reality experience was found to be an effective medium by which to teach marine science and the consequences of climate change.

Web based tools

Interactive, web-based data visualizations are important for making climate change information more accessible (Lumley et al., 2022). Web-based visualizations have various applications, such as sea level visualization tools, climate change adaptation tools, maps, and climate change websites. The usage of these applications can vary according to the target audience, and the context in which the climate issue is being addressed. For example, Neset et al. (2016) examined climate change adaptation through web mapping, while Lee (2018) aggregated data from scientific organizations through visualization of climate change news. Hewitson et al. (2017) tried combining websites that provide climate change information under a single roof. Fish (2020) developed a climate change map addressing variables such as temperature and rising sea levels.

Accessible *web 2.0* tools empower teachers to effectively educate students on climate change and its impacts. *Climate Trace* (https://climatetrace.org/map) is a practical website for monitoring the amount of anthropogenic greenhouse gas emissions worldwide, *Earth.nullschool.net* (https://earth.nullschool.net/) can access current weather, ocean, and pollution conditions, while *Climate Reanalyzer* (https://climatereanalyzer.org/) is able to provide climate change data.

Digital games

There are a variety of digital games and simulations focusing on climate change processes, the role of human systems, and potential impacts and interventions (Wu and Lee, 2015). Digital games can engage all age groups and provide first-hand experiential learning experiences (Squire, 2006). In this respect, climate change games are considered *serious games* (Abt, 1987). Research suggests game-based learning can lead to cognitive gains and produce emotional and motivational outcomes (Bellotti et al., 2013; Vogel et al., 2006). Climate change, commonly model increasing CO^2 levels in the atmosphere and other equally well-known topics, such as melting glaciers (Robinson and Ausubel, 1983). Over time, games related to other aspects of climate change have emerged.

Gustafsson et al. (2010) designed *PowerAgent*, a game focused on reducing power consumption through actions like adjusting household heating and turning off standby appliances. PowerAgent uses real power consumption data from home metering devices. Launched in 2013, the *Habitat* game encourages users of all ages to take environmentally-friendly measures, rewarding them with a badge for doing so. Lee et al. (2013) created the game *GREENIFY*. The game raised awareness about climate change and attempted to foster positive attitudes towards learning more about the subject. Ro et al. (2017) developed a team-based game called *Cool Choices,* in which players must reduce the amount of energy they consume. Reportedly, the game, which awards points based on eco-friendly actions, led to long-term reductions in electricity consumption among people who were large consumers.

Digital games have diverse structures. Role-play simulations (RPS), categorized as serious games, assign participants individual roles through which they participate in mock negotiations and decision-making activities that reflect real-world scientific and institutional realities (Gordon et al., 2011). RPSs are widely used in educational contexts because they promote emergency preparedness and personal deliberation (Yee and Bailenson, 2006). They also support the development of empathy by encouraging participants to interact openly with those of different perspectives (Schenk and Susskind, 2015). Research has shown that RPSs can be used to understand the potential impacts of climate change and combat scientific uncertainty (Susskind et al., 2015). The study conducted by Rumore et al. (2016) can be cited as an example of a good application of RPSs. Their research found that the use of RPSs in teaching climate change increases collaborative capacity, facilitates climate change adaptation, and provides an enriched understanding of the subject complexity and what adaptation means in practice forthe participants. When designing games, it is important to consider players' ages (Douglas and Brauer, 2021). In the game *Energy Cat*, it was observed that no long-term energy consumption occurred amongst players. Participants reported that they were not provided clear instructions and that the game seemed too childish and not enjoyable (Casals et al., 2020).

Video games

Video games differ from digital games in that they run on video software and often do not require an internet connection. Not needing an internet connection makes them more accessible (Barreteau et al., 2007). Climate change video games can raise awareness of climate issues rising and inculcate positive attitudes (Mathevet et al., 2007). The widespread popularity of video games speaks to their potential impact (indiecade, 2022).

Ho et al. (2022) examined the impact of *Nintendo Animal Crossing: New Horizon* (ACNH) on users' pro-environmental behavior. In the game, players are required to grow tree crops and feed fish and insects on a virtual island to make a living. In addition, from an economic point of view, players need to take care of the fruits and spend limited amounts of wood. For this, they need to protect the trees and develop a strategy for sustainability. If trees are cut down in the game, they will not grow back. Flowers, however, can grow indefinitely without intervention. Therefore, the game encourages players to plant flowers. Barcena-Vazquez et al. (2023) prepared a video game to support global warming awareness of museum visitors. "From Pole to Pole: Climate Change and Global Warming, Reto Global" was tested with the support of 30 participants aged 13–21. Pre-test-post-test experimental design was applied, and both participant perceptions about the museum and the change in their global warming awareness were measured and examined. The game commenced with a videodetailing actions to combat climate change and presentinga panoramic view of a polluted, devastated Earth. A main character then appears on screen to scientifically explain why theplanet looks the way it does, and then calls on the player to play several mini-games to restore the Earth. The study states that the video, including three mini games, was effective in raising awareness amongst museum staff and visitors.

Mobile apps

Mobile applications enable independent learning and direct access to information by providing access from anywhere. (Magne Vestøl, 2011). Mobile learning expands access to knowledge by enabling unguided learning (Vosloo, 2012). Research conducted in the last decade suggests that mobile applications can impact student comprehension, motivation, and learning satisfaction (Wang, 2017). As such, mobile applications (EnergySmart, Vodoo Skies Normal or Not, Chasing Ice, Commute Greener, AllertMe Energy Map, PowerPup, EnergySmart, Vodoo Skies Normal or Not, Chasing Ice, Commute Greener, AllertMe Energy Map, PowerPup), are used to raise awareness about climate change and its effects by making accessible useful information ranging from pollution maps to information on how to clean the oceans. Moreover, these apps are supported by iOS, Android, Windows, and MacOSX. Easy to both access and use, mobile applications have been the subject of many studies.

Sullivan et al. (2016) used mobile applications to determine the cost of personal carbon emissions and to show the impact of this cost on health expenditures. In addition, Wallace and Bodzin (2017) reported that mobile applications increased student interest in science and technology.

Conclusion

Climate change is one of the most critical challenges facing humanity, and it requires global solutions. Technology provides solutions to overcome this challenge and adopt sustainable practices. Educating people on climate change and raising public awareness is possible through various applications of modern technology. Many positive outcomes have been demonstrated by researchers studying technological applications in climate change education. Many studies have reported that higher-order thinking, problem-solving, collaboration, and understanding of abstract concepts have been improved with such integration. It is hoped that this chapter will be useful for researchers who will conduct studies in this field in the future to know the technologies commonly used in the field of climate change and to evaluate the outputs obtained.

References

Abt, C.C. (1987). *Serious games.* University press of America.

Achieve (2013). The next generation science standards. Available at http://www.nextgenscience.org/next-generation-science-standards

Barcena-Vazquez, J., Caro, K., Bermudez, K. and Zatarain-Aceves, H. (2023). Designing and evaluating Reto Global, a serious video game for supporting global warming awareness. *International Journal of Human-Computer Studies*, *177*, 103080. https://doi.org/10.1016/j.ijhcs.2023.103080

Barreteau, O., Le, Page, C. and Perez, P. (2007). Contribution of simulation and gaming to natural resource management issues: An introduction. *Simul Gaming, 38*(2), 185–194. https://doi.org/10.1177/1046878107300660

Bell, T., Urhahne, D., Schanze, S. and Ploetzner, R. (2010). Collaborative inquiry learning: Models, tools, and challenges. *International Journal of Science Education*, *32*(3), 349–377. https://doi.org/10.1080/09500690802582241

Bellotti, F., Kapralos, B., Lee, K., Moreno-Ger, P. and Berta, R. (2013). Assessment in and of serious games: An overview. *Advances in Human-Computer Interaction, 2013*, 1–1. https://doi.org/10.1155/2013/136864

Bodzin, A.M., Fu, Q., Peffer, T.E. and Kulo, V. (2013). Developing energy literacy in US middle-level students using the geospatial curriculum approach. *International*

Journal of Science Education, *35*(9), 1561–1589. http://dx.doi.org/10.1080/09500693.2013.769139

Bush, D., Sieber, R., Seiler, G. and Chandler, M. (2018). Examining educational climate change technology: how group inquiry work with realistic scientific technology alters classroom learning. *Journal of Science Education and Technology*, *27*, 147–164. http://dx.doi.org/10.1007/s10956-017-9714-0

Campbell, T. and Oh, P.S. (2015). Engaging students in modeling as an epistemic practice of science: An introduction to the special issue of the Journal of Science Education and Technology. *Journal of Science Education and Technology*, *24*, 125–131. http://dx.doi.org/10.1007/s10956-014-9544-2

Casals, M., Gangolells, M., Macarulla, M., Forcada, N., Fuertes, A. and Jones, R. V. (2020). Assessing the effectiveness of gamification in reducing domestic energy consumption: Lessons learned from the EnerGAware project. *Energy and Buildings*, *210*, 109753. http://dx.doi.org/10.1016/j.enbuild.2019.109753

Cheng, G., Han, J. and Lu, X. (2017). Remote sensing image scene classification: Benchmark and state of the art. *Proceedings of the IEEE*, *105*(10), 1865–1883.

Cox, H., Kelly, K. and L. Yetter (2014). Using remote sensing geospatial technology for climate change education. *Journal of Geoscience Education*, *62*(4), 609–620. http://dx.doi.org/10.5408/13-040.1

Dede, C.J., Jacobson, J. and Richards, J. (2017). Introduction: Virtual, augmented, and mixed realities in education. pp. 1–16. *In*: Liu, D., Dede, C., Huang, R. and Richards, J. (eds.). *Virtual, augmented, and mixed realities in education*; smart computing and intelligence, Springer.

Dimitrovski, I., Kitanovski, I., Kocev, D. and Simidjievski, N. (2023a). Current trends in deep learning for Earth Observation: An open-source benchmark arena for image classification. *ISPRS Journal of Photogrammetry and Remote Sensing*, *197*, 18–35. http://dx.doi.org/10.1016/j.isprsjprs.2023.01.014

Dimitrovski, I., Kitanovski, I., Panov, P., Kostovska, A., Simidjievski, N. and Kocev, D. (2023b). Aitlas: Artificial intelligence toolbox for earth observation. *Remote Sensing*, *15*(9), 2343. http://dx.doi.org/10.3390/rs15092343

Douglas, B.D. and Brauer, M. (2021). Gamification to prevent climate change: A review of games and apps for sustainability. *Current Opinion in Psychology*, *42*, 89–94. http://dx.doi.org/10.1016/j.copsyc.2021.04.008

Fish, C.S. (2020). Cartographic content analysis of compelling climate change communication. *Cartography and Geographic Information Science*, *47*(6): 492–507. http://dx.doi.org/10.1080/15230406.2020.1774421

Game4Sustainability. (n.d). Gamepedia. https://games4sustainability.org/gamepedia/

Gautier, C. and Solomon, R. (2005). A preliminary study of students asking quantitative scientific questions for inquiry-based climate model experiments. *Journal of Geoscience Education*, *53*(4), 432–443.

Gordon, E., Schirra, S. and Hollander, J. (2011). Immersive planning: A conceptual model for designing public participation with new technologies. *Environment and Planning B: Planning and Design*, *38*(3), 505–519. http://dx.doi.org/10.1068/b37013

Gustafsson, A., Katzeff, C. and Bang, M. (2010). Evaluation of a pervasive game for domestic energy engagement among teenagers. *Computers in Entertainment (CIE)*, *7*(4), 1–19. http://dx.doi.org/10.1145/1658866.1658873

Hestness, E., McDonald, R.C., Breslyn, W., McGinnis, J.R. and Mouza, C. (2014). Science teacher professional development in climate change education informed by the Next

Generation Science Standards. *Journal of Geoscience Education*, *62*(3), 319–329. http://dx.doi.org/10.5408/13-049.1

Hewitson, B., Waagsaether, K., Wohland, J., Kloppers, K. and Kara, T. (2017). Climate information websites: an evolving landscape. *Wiley Interdisciplinary Reviews: Climate Change*, *8*(5), e470. http://dx.doi.org/10.1002/wcc.470

Ho, M.T., Nguyen, T.H.T., Nguyen, M.H., La, V.P. and Vuong, Q.H. (2022). Virtual tree, real impact: how simulated worlds associate with the perception of limited resources. *Humanities and Social Sciences Communications*, *9*(1), 1–12. http://dx.doi.org/10.1057/s41599-022-01225-1

indiecade. (2022). Climate Jam. Retrieved from https://www.indiecade.com/climate-jam/

Jain, H., Dhupper, R., Shrivastava, A., Kumar, D. and Kumari, M. (2023). AI-enabled strategies for climate change adaptation: Protecting communities, infrastructure, and businesses from the impacts of climate change. *Computational Urban Science*, *3*(1), 25–35. http://dx.doi.org/10.1007/s43762-023-00100-2

Kulp, S.A. and Strauss, B.H. (2018). CoastalDEM: A global coastal digital elevation model improved from SRTM using a neural network. *Remote Sensing of Environment*, *206*, 231–239.

Lee, E. (2018). *Stories in the data: An analysis of climate change visualisations in online news* (Doctoral dissertation). Sidney University.

Lee, E.A.L., Wong, K.W. and Fung, C.C. (2010). How does desktop virtual reality enhance learning outcomes? A structural equation modeling approach. *Computers & Education*, *55*(4), 1424–1442. http://dx.doi.org/10.1016/j.compedu.2010.06.006

Lee, J.J., Ceyhan, P., Jordan-Cooley, W. and Sung, W. (2013). GREENIFY: A real-world action game for climate change education. *Simulation & Gaming*, *44*(2–3), 349–365. http://dx.doi.org/10.1177/1046878112470539

Lumley, S., Sieber, R. and Roth, R. (2022). A framework and comparative analysis of web-based climate change visualization tools. *Computers & Graphics*, *103*, 19–30. http://dx.doi.org/10.1016/j.cag.2021.12.007

Magne Vestøl, J. (2011). Digital tools and educational designs in Norwegian textbooks of religious and moral education. *Nordic Journal of Digital Literacy*, *6*(1–2), 75–88. http://dx.doi.org/10.18261/issn1891-943x-2011-01-02-06

Makransky, G. and Petersen, G.B. (2019). Investigating the process of learning with desktop virtual reality: A structural equation modeling approach. *Computers & Education*, *134*, 15–30. http://dx.doi.org/10.1016/j.compedu.2019.02.002

Markowitz, D.M., Laha, R., Perone, B.P., Pea, R.D. and Bailenson, J.N. (2018). Immersive virtual reality field trips facilitate learning about climate change. *Frontiers in Psychology*, *9*, 2364. http://dx.doi.org/10.3389/fpsyg.2018.02364

Mathevet, R., Le Page, C., Etienne, M., Lefebvre, G., Poulin, B., Gigot, G., Proréol, S. and Mauchamp, A. (2007) BUTORSTAR: A role-playing game for collective awareness of wise reedbed use. *Simul Gaming* *38*(2), 233–262. http://dx.doi.org/10.1177/1046878107300665

Monroe, M.C., Plate, R.R., Oxarart, A., Bowers, A. and Chaves, W.A. (2019). Identifying effective climate change education strategies: A systematic review of the research. *Environmental Education Research*, *25*(6), 791–812. http://dx.doi.org/10.1080/13504622.2017.1360842

Neset, T.S., Opach, T., Lion, P., Lilja, A. and Johansson, J. (2016). Map-based web tools supporting climate change adaptation. *The Professional Geographer*, *68*(1), 103–114. http://dx.doi.org/10.1080/00330124.2015.1033670

Ou, K.L., Chu, S.T. and Tarng, W. (2021). Development of a virtual wetland ecological system using VR 360 panoramic technology for environmental education. *Land, 10*(8), 829–933. http://dx.doi.org/10.3390/land10080829

Ro, M., Brauer, M., Kuntz, K., Shukla, R. and Bensch, I. (2017). Making Cool Choices for sustainability: Testing the effectiveness of a game-based approach to promoting pro-environmental behaviors. *Journal of Environmental Psychology, 53*, 20–30. http://dx.doi.org/10.1016/j.jenvp.2017.06.007

Robinson, J. and Ausubel, J.H. (1983). A game framework for scenario generation for the CO^2 issue. *Simulation & Games, 14*(3), 317–344. http://dx.doi.org/10.1177/104687818301400306

Ruan, B. (2022). VR-assisted environmental education for undergraduates. *Advances in Multimedia, 2022*, 1–8.

Rumore, D., Schenk, T. and Susskind, L. (2016). Role-play simulations for climate change adaptation education and engagement. *Nature Climate Change, 6*(8), 745–750. http://dx.doi.org/10.1038/nclimate3084

Schenk, T. and Susskind, L. (2015). *Action research for climate change adaptation: Developing and applying knowledge for governance.* Routledge.

Schott, C. and Marshall, S. (2018). Virtual reality and situated experiential education: A conceptualization and exploratory trial. *Journal of Computer Assisted Learning, 34*(6), 843–852. http://dx.doi.org/10.1111/jcal.12293

Squire, K. (2006). From content to context: Videogames as designed experience. *Educational Researcher, 35*(8), 19–29. http://dx.doi.org/10.3102/0013189x035008019

Sullivan, R.K., Marsh, S., Halvarsson, J., Holdsworth, M., Waterlander, W., Poelman, M.P. and Maddison, R. (2016). Smartphone apps for measuring human health and climate change co-benefits: A comparison and quality rating of available apps. *JMIR mHealth and uHealth, 4*(4), e5931. http://dx.doi.org/10.2196/mhealth.5931

Susskind, L., Rumore, D., Hulet, C. and Field, P. (2015). *Managing climate risks in coastal communities: Strategies for engagement, readiness and adaptation.* Anthem Press.

Swarat, S., Ortony, A. and Revelle, W. (2012). Activity matters: Understanding student interest in school science. *Journal of Research in Science Teaching, 49*(4), 515–537. http://dx.doi.org/10.1002/tea.21010

Vogel, J.J., Vogel, D.S., Cannon-Bowers, J., Bowers, C.A., Muse, K. and Wright, M. (2006). Computer gaming and interactive simulations for learning: A meta-analysis. *Journal of Educational Computing Research, 34*(3), 229–243. http://dx.doi.org/10.2190/flhv-k4wa-wpvq-h0ym

Vosloo, S. (2012). UNESCO policy guidelines for mobile learning. *URL:https://unesdoc. unesco. org/ark:/48223/pf0000219641*

Wallace, D. and Bodzin, A. (2017). Developing scientific citizenship identity using mobile learning and authentic practice. *The Electronic Journal for Research in Science & Mathematics Education, 21*(6), 1–26.

Wang, Y. H. (2017). Integrating self-paced mobile learning into language instruction: Impact on reading comprehension and learner satisfaction. *Interactive Learning Environments, 25*(3), 397–411. http://dx.doi.org/10.1080/10494820.2015.1131170

Wibeck, V. (2014). Enhancing learning, communication and public engagement about climate change–some lessons from recent literature. *Environmental Education Research, 20*(3), 387–411. http://dx.doi.org/10.1080/13504622.2013.812720

World Economic Forum (2022). *How information tech can address challenges in climate change and education*. Avaliable on https://www.weforum.org/agenda/2022/09/how-ict-can-address-challenges-climate-change-education/

Wu, J.S. and Lee, J.J. (2015). Climate change games as tools for education and engagement. *Nature Climate Change*, *5*(5), 413–418. http://dx.doi.org/10.1038/nclimate2566

Xiong, Z., Zhang, F., Wang, Y., Shi, Y. and Zhu, X.X. (2022). Earthnets: Empowering ai in earth observation. *arXiv preprint arXiv:2210.04936*.

Yee, N. and Bailenson, J.N. (2006). Walk a mile in digital shoes: The impact of embodied perspective-taking on the reduction of negative stereotyping in immersive virtual environments. *9th Annu. Int. Work. Presence International Society for Presence Research.*

Chapter 4

Digital Technologies for Climate Education

A Scoping Review of Empirical Studies

Paulina Rutecka,[1,*] *Karina Cicha,*[2] *Mariia Rizun*[1] and *Artur Strzelecki*[1]

Introduction

Since the signing of the Kyoto Protocol in 1997, the countries of the European Union have made an intensified effort to prevent climate changes caused by industrial development. The EU and international organizations such as the United Nations have undertaken several major ecological and sustainable initiatives in the last two decades. Such initiatives as the Green Climate Fund (2010), the UN's 2030 Agenda of 17 Sustainable Development Goals (SDGs), and the Green Deal strategy (2019) directly aim to improve the climate situation globally. However, in addition to large-scale solutions, it is equally essential to educate

[1] Department of Informatics, University of Economics in Katowice, Katowice, Poland.
Email: mariia.rizun@ue.katowice.pl
artur.strzelecki@ue.katowice.pl

[2] Department of Communication Design and Analysis, University of Economics in Katowice, Katowice, Poland.
Email: karina.cicha@ue.katowice.pl

* Corresponding author: paulina.rutecka@ue.katowice.pl

people on climate change, ecological issues, and their countermeasures. Climate change impacts all aspects of human life (Ofori et al., 2023). Awareness of its effects and consequences, both potential and already occurring, seems crucial not only in the broad debate on the environment but also in the sphere of education. This necessity is a challenge for modern education institutions. The challenge lies in how environmental education can and should be conducted.

Researchers have questioned the necessity of using technology in environmental teaching and learning (Greenwood and Hougham, 2015), naming the fact that computer-mediated technologies tend to distort the human-environment relationship as a reason for that (Bowers, 2006). Furthermore, some educational institutions' implementations of new technologies are negatively labeled *technosolutionism* or are even considered a form of greenwashing, i.e., the deceptive presentation of exaggerated or false claims about these institutions' environmental practices (Stein, 2023). However, many studies show not only the necessity of environmental education but also the efforts undertaken, particularly by higher educational institutions (HEIs) worldwide, to teach about climate change and environmental issues (Li and Liu, 2022; Xabregas and Brasileiro, 2023).

HEIs Higher educational institutions play a vital role in climate-related education because they not only build knowledge through research but also provide possible solutions, thereby equipping both current and future leaders with the tools necessary to confront environmental challenges (Leal Filho et al., 2023). However, problems with climate education at universities do exist and need to be addressed. As stated previously, universities are responsible for educating on climate change, thereby building students' environmental awareness and allowing them to recognize environmental processes and problems, all of which leads to pro-environmental behavior and protective ecological behavior in everyday life (Kousar et al., 2022; Yeung, 1998).

In studies focused on educational process it is mentioned that university courses related to health and engineering do not cover climate change issues, even though these areas are closely linked to the environment (Axelithioti et al., 2023; Palmeiro-Silva et al., 2021). Nevertheless, research indicates that students are aware of the importance of these problems, even if they are not part of curricula. Students recognize human interference with nature as the main cause of climate change (Nadeem and Nawaz, 2023).

In studies focusing on teachers in connection with environmental issues, it was observed that even though some teachers had general knowledge about climate change, they often had an ambiguous or wrong understanding of the concepts of climate change, global climate warming, greenhouse effects, and the interrelatedness of these issues (Wan et al., 2023). This shows the necessity of conducting a broader information campaign about environmental issues that targets teachers.

Learning and teaching about climate change is complex. Studies suggest that teacher education is the first challenge in implementing effective climate change education. Research states that teachers must develop an extensive knowledge base in order to design and carry out effective climate education (Favier et al., 2021), and such efforts are necessary at every level of education. The second challenge in introducing environmental education into didactic practice is that it requires an interdisciplinary approach, whereas current research clearly shows that it is inherently a multidisciplinary endeavor (Mohan et al., 2023) that requires a broad spectrum of competencies. Studies show that international policies for sustainability education are expected to be introduced in educational processes; however, barriers to doing so can be observed both at the level of curricula and in the education system as a whole (Parry and Metzger, 2023).

As for the means of building students' awareness, different positions are taken by researchers. There is a worldwide discussion regarding sustainable pedagogy at higher-education institutions, but the nature of the content that should be included in climate change education (Fuertes-Camacho et al., 2019) and how it should be conveyed (Seatter and Ceulemans, 2017) is well established. The Burns Model of Sustainability Pedagogy introduced a set of elements to be included in university courses: ecological design, systemic and interdisciplinary learning, active and engaged learning processes, and attention to place-based learning (Burns, 2009). It is clear that sustainability education at higher levels of education requires varied pedagogical approaches so that students may gain broad experience of environmental issues through methods such as problem-based learning and experiential learning (Missimer and Connell, 2012). Also, action-oriented learning processes have been shown to foster thinking across disciplines (Loeber et al., 2007). Such approaches are consistent with the constructivist learning perspective, in which students are challenged to develop responses to defined problems, eventually deriving solutions through case studies and active participation involving brainstorming, dialogue, and teamwork (Seatter and Ceulemans, 2017). More profound engagement is also possible through critical self-reflection (Elder et al., 2023).

Another aspect considered in many studies on climate change education is the use of information and communication technology (ICT) in environmental education. The urgency of climate change and the rapid development of ICT both represent a challenge for higher education institutions (HEIs) as they are forced to reconsider their traditional ways of teaching (Versteijlen and Wals, 2023). For example, teaching formats such as webinars could reduce the carbon footprint of students and staff. Researchers also mention some activities undertaken by university students or teachers that involve forms of transport that generate environmental problems and are considered to have a social impact on the environment (Baer, 2023; Shields and Lu, 2023). Online education tools, such as online classrooms and tutorials, can provide significant advantages, such as reduced need for infrastructure and reduced carbon emissions (Alla and Chen, 2017). However, researchers are aware that certain obstacles must be overcome when introducing ICT in environmental education: structural barriers, i.e., lack of support and incentives for interdisciplinary teaching and community-based research; cultural barriers, understood as biases towards specific disciplines, or lack of experience and knowledge about interdisciplinary or experiential teaching; and, finally, financial barriers, namely insufficient resources (Wade et al., 2020).

Despite the complexity of climate education and the myriad opinions regarding its effective implementation, awareness of the gravity of climate-related issues prompted us to look for examples of digital tools used in climate education at the higher education level. The research questions for this chapter are as follows:

RQ1: Does climate education use digital tools in higher-education teaching processes?

RQ2: Which digital tools are implemented in climate education in higher education?

The answers to these research questions allowed us to achieve the objective of our paper: to reveal the digital tools used in higher-education climate education that have been presented in the literature in the last five years.

Methods and Materials

This section describes the sequential phases of the analysis undertaken. Given the broad thematic spectrum encompassed by this investigation, the scoping review methodology was used, incorporating the initial five stages of the methodological framework articulated by Arksey and O'Malley (2005), with subsequent refinements of Levac (2010). While initially tailored for use in medical science, the scoping review framework, which encompasses educational tools and methodologies, has also been used in educational research (Adnan and Xiao, 2023; Jaleniauskiene and Kasperiuniene, 2023; Sormunen et al., 2022).

The scoping review was conducted to investigate the implementation of tools and teaching methods in teaching about climate change at higher-education institutions. According to Arksey and O'Malley's (2005) methodological framework, the research steps can be described as follows:

- formulation of research questions,
- identification of appropriate academic works,
- selection of pertinent studies,
- systematic charting of collated data,
- compilation and explication of ascertained outcomes.

HEIs' didactic processes, focusing on climate change as the topic of courses in these institutions. The next step was to find scientific papers relevant to the topic which had been published in the last five years (2019 to 2023) by searching the Scopus and Web of Science databases. We decided to use these two databases because they are the most extensive abstract and citation databases for academic literature. We did not search for papers in the Google Scholar database because although this database covers every document which contains the defined keywords, it also includes works that are not necessarily scientific or peer-reviewed. Combinations of the following search terms and subheadings were considered appropriate for the conducted study: "Climate change", "Climate education", "Climate AND education", "Climate change AND education", and "higher education" or "HEI". Quantitative search results for the defined key phrases are presented in Table 1.

Next, we defined inclusion and exclusion criteria to limit the resources found. We restricted the original research papers (published between January 2019 and November 2023) to those written in English that describe tools (including digital tools) and methods used in higher-

Table 1: Key search phrases and search results in Scopus and Web of Science databases.

	Scopus	Web of Science
"Climate change" AND "higher education"	530	493
"Climate education" AND "higher education"	11	8
"Climate change" AND education AND hei	27	15
"Climate" AND education AND hei"	45	19
Total:	613	535

education teaching. We did not use any further exclusions regarding, for instance, study type (e.g., book chapters or editorials), or methodologies (e.g., expert reviews, systematic reviews, scoping reviews, and narrative reviews). While searching for relevant papers that would later be used for the data extraction, we first removed papers that appeared in both databases. At that point, the initial number of 1148 papers was reduced to 766. The next step was title scanning. For further analysis, we agreed that a paper title should include a type of tool or method used in climate change education, and that this information should be combined with the field of study in which this tool or method was implemented. We eliminated all papers with unrelated titles. The number of papers remaining after title scanning was 320.

The next step was analysis of the papers' abstracts. We asked such questions as: Is this paper relevant to this study? Does it focus on higher education? Does it present information on teaching tools and methods? Is this study original? We eliminated all papers showing any form of a literature review. This reduced the total number of papers qualified for inclusion to 113. The last stage of elimination focused on access to the full papers. Since not every paper was accessible as a full text in the online databases, we established the final number of 61 papers in the study.

During the analysis of the included publications, Cicha and Rutecka's (2023) catalog of methods and digital tools used in higher education was utilized. This catalog identifies 29 categories of tools and methods applied in teaching and learning. During this review, one or more categories from the list were assigned to the publications analyzed in this study. Not all categories from the previous catalog could be assigned. In the publications that qualified for the study, we found only 18 methods and tools of modern digital education of the 29 identified in Cicha and Rutecka's catalog (2023). Some publications focus generally on frameworks without specifying the exact tools used or on frameworks

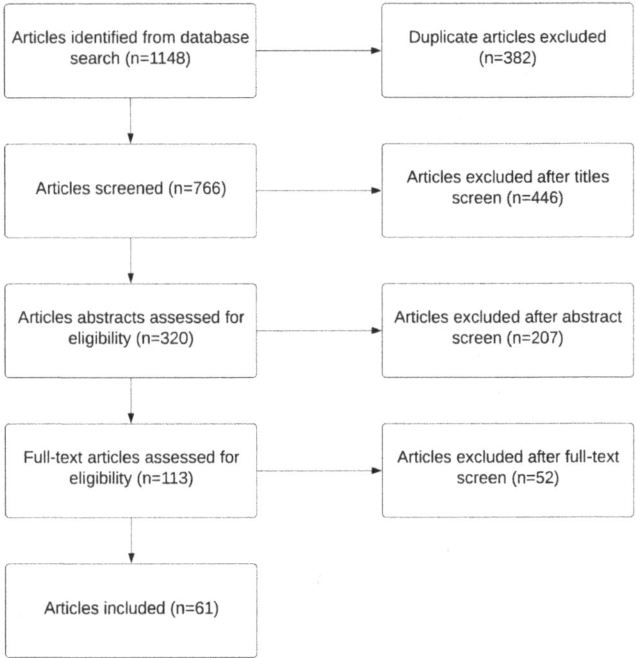

Figure 1: Steps in the elimination of papers in the conducted research.

that do not employ any tools. Such frameworks include COIL (Collaborative Online International Learning) and CLEWs (Climate, Land-use, Energy and Water Systems), which were either combined in a single category or included in another adequate category.

Results

The first result obtained during the scoping review was quantitative information about the number of papers in which authors indicated the type of digital tool and the scope of its use in climate education. Unfortunately, considering that the scoping review included publications from a period of five years, the number of papers describing the usage of digital tools in environmental education is low (only 61). When trying to categorize digital tools used in climate change education, we noticed that there are 17 specific categories and an "Other" category that contains tools that do not fit into any other category. The categories used to assign digital climate education tools in higher education refer to the previously conducted study on the use of digital techniques in higher education (Cicha and Rutecka, 2023). Figure 2 shows the categories and the number of studies that reported their use in climate education.

Within the methods listed in the catalog, the most frequently used is the game-based approach, especially with serious games designed for education purposes. Game-based learning is about developing new concepts and skills through digital and non-digital games (Adipat et al., 2021). This method is considered advantageous in increasing students' motivation and engagement (Adipat et al., 2021), teamwork and team building (Dichev and Dicheva, 2017), and risk-taking and experimentation (Martí-Parreño et al., 2016). In the analyzed studies, game-based teaching built on, for example, role-playing environments, was pointed out as beneficial for students in terms of decision-making (Stoeth and Carter, 2023), familiarizing students with the complex interactive characteristics of such systems (Thompson et al., 2022), and increasing students' engagement concerning climate change-related issues (Vázquez-Vílchez et al., 2021).

Video communication refers to tools for real-time audiovisual transmission. As for the use of video communication in climate education, the possibility of increasing internationalization by exchanging experiences and views on climate change in a global environment is pointed out as an advantage (Falkenberg and Joyce, 2023). Other advantages include expanding students' knowledge on environmental issues (Straßer et al., 2023) and more efficient access to and use of up-to-date information (Baptista et al., 2021). It is worth mentioning that many

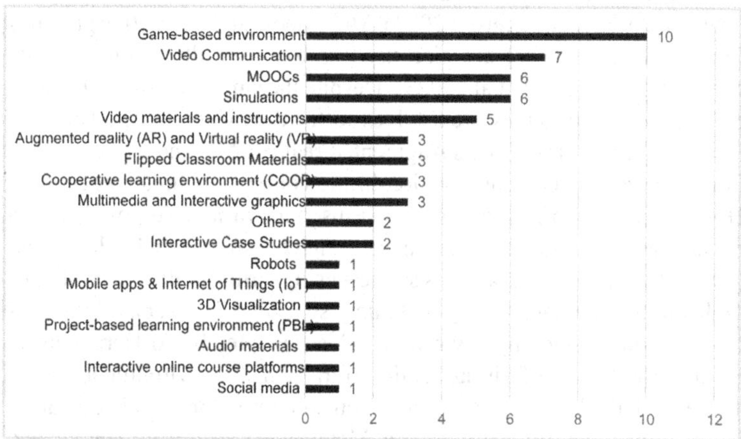

Figure 2: Digital tools for sustainable education.

of the analyzed studies were conducted during the COVID-19 pandemic, therefore the use of certain ICT solutions was not a carefully planned educational choice but was forced by the situation in many cases. Nevertheless, students' use of video communication tools allows them to avoid carbon-intensive transport and overcome barriers in research participation (Elder et al., 2023).

An interesting form of teaching about climate change is simulation. By using specially designed tools in, for instance, architecture studies, students can improve their building designs by simulating their environmental costs (de Gaulmyn and Dupre, 2019). Also in engineering, active learning is supported by simulations of energy management platforms for smart and green building design (Apichayakul et al., 2020). Technological progress has allowed universities to introduce Massive Open Online Courses (MOOCs) in teaching practice. This technology is considered one of the most engaging for students (Senevirathne et al., 2022) and it has been demonstrated that implementing MOOCs in environmental education can support networking development (Senevirathne et al., 2021) and attract participants within low-resource contexts (Barteit et al., 2019).

The use of video-based materials and tutorials in the climate change teaching process (indicated five times in the studied papers) refers, for instance, to implementing environmental films paired with viewer-response activities such as reflections and discussions to create emotional engagement (Esmail and Matthews-Roper, 2022). Video materials have also been used as additional elements of broad educational projects (Membrillo-Hernández et al., 2023). Multimedia and interactive graphics (indicated three times) are also used as tools for illustrating climate change issues (Cotton et al., 2023). Despite the simplicity of their use, in the analyzed studies they are used less often than video materials.

Using tools classified as Augmented reality (AR) and Virtual reality (VR) is a very interesting solution in the context of climate education. However, although this solution can be used for various purposes, it does not appear often enough in the papers we have analyzed. In Pavlova et al. (2020), for example, the authors suggest using virtual reality technologies to learn foreign languages, including specialized vocabulary that is helpful in understanding environmental issues. Membrillo-Hernández et al. propose the use of virtual reality to transfer the environment known as the Global Classroom, i.e., classes on an international scale operating in an online environment, to the Metaverse (Membrillo-Hernández, Cuervo-Bejarano, Mejía-Manzano, et al., 2023), and geoscientists have proposed learning with the AR/VR-based "GeoTrails" tool, which offers students virtual field trips (Maloney et al., 2023).

Flipped classroom materials are educational materials, such as previously recorded video lectures, that are prepared and made available to students electronically. This method of delivering materials minimizes waste and reduces the carbon footprint associated with printing materials (Mulla and Ratnayake, 2020). Thanks to the fact that materials are made available to students before face-to-face classes, they have time to become familiar with the material and can implement active learning strategies for the classroom (Tomas et al., 2019). This approach increases students' engagement, and they perform better and demonstrate increased awareness of climate issues (Jeong et al., 2021).

Studies have noted that students' engagement was higher when they carried out projects collaboratively or created teams that could compete with each other. One example of a cooperative learning environment (COOP) was a board game in a virtual space that students played in teams (Vázquez-Vílchez et al., 2021). This approach was also used in an international educational project that involved seven European universities (De Stefani and Han, 2022).

Other methods mentioned in the study included Interactive Case Studies (indicated 2 times), Social media (1), Interactive online course platforms (1), Audio materials (1), 3D Visualization (1), Mobile apps & Internet of Things (1) and Robots (1). Project-based learning environments were also rarely mentioned (1), but Problem-based Learning (PbBL) was mentioned six times. In the analysis, project-based learning and problem-based learning were classified as variations of the Challenge-based learning (CBL) approach, which appeared in 17 studies, but the primary form of conducting educational activities for students was not indicated. The advantage of the challenge-based approach is students' involvement in designing solutions for real environmental and social problems. Two publications describe the use of the COIL method, which is dedicated to teaching about climate change problems in an online environment and is based on cooperation between groups of students from universities in different countries (Membrillo-Hernández, Cuervo-Bejarano and Vázquez-Villegas, 2023; Membrillo-Hernández, Cuervo-Bejarano, Mejía-Manzano et al., 2023). One publication describes a method of working with students using online tools to carry out a sustainability audit (Emblen-Perry, 2019); another describes the use of eye-tracking, (Södervik and Vilppu, 2021). These tools were not previously included in the catalog (Cicha and Rutecka, 2023).

Discussion

The methods found in the papers we analyzed primarily suggest that teaching should be associated with challenges and increase student engagement in facing real-world problems (Gregory and Lewin, 2023). The described methods include student-centered learning (Van Heuvelen et al., 2020), active learning (Bartlett et al., 2022; Emblen-Perry, 2019; Leichenko and O'Brien, 2020), experiential learning also called learning by doing (Elder et al., 2023; Wade et al., 2020), and collaborative learning (Capetola et al., 2022; Versteijlen and Wals, 2023). The studies also emphasize the importance of interdisciplinarity as it gives students a broad perspective (Capetola et al., 2022; Wade et al., 2020). Essential competencies in the field of solving climate problems include effective communication skills (Wade et al., 2020), which can be successfully developed thanks to digital tools. In this case, digital tools can also help in international communication and strengthen cooperation between students of different cultures, thus helping them to discover other points of view. Unfortunately, according to the analyzed publications, ICT tools are often not utilized for this purpose. Among the publications that qualified for the study, some focused on Challenge-based Learning (CBL) or indicated that digital tools or online tools had been used to implement teaching under the CBL model. CBL uses a mix of basic digital tools, such as videos or online communication.

Several recurring themes were observed in the publications for which we have analyzed the full text but which we did not ultimately include in the study because they did not specify a particular teaching method or tool. These studies primarily focused on the carbon footprint of international travel and of commuting to on-campus classes (Versteijlen and Wals, 2023). It was also frequently noted that climate change and sustainable development issues are not sufficiently addressed in study programs and course curricula, and that there are discrepancies in students' climate change awareness depending on a university's location or students' gender, age, or study program.

The research by Versteijlen and Wals (2023) was generally dedicated to sustainability-oriented blended learning; these authors analyzed 38 papers to determine the methods by which blended learning is introduced into the education process. They revealed various types of flipped classroom learning with the usage of online discussions and quizzes, physical and virtual labs, video lectures, interactive online textbooks, gamification, etc. Although the topics (e.g., Project management, English,

ICT, Medical science) of the courses that applied blended learning were not directly connected to climate change, through the use of online or blended learning each of these courses reduced negative effects on the environment by allowing the students not to travel to their HEIs to study.

Although they are not based on digital tools, two of the most interesting and frequently mentioned approaches to teaching about climate change and sustainable development that were revealed in the analyzed publications are Arts-based approaches and Living Lab. The latter was mentioned in the rejected studies as many as six times. Universities can reflect society on a micro scale, thus they are an excellent field for conducting research and testing innovations as a "living laboratory" (Martek et al., 2022). As the research shows, this approach is currently implemented on a small scale and often fragmentarily, but researchers postulate that this state should be changed.

Crosling et al. (2020) explored academic university staff's knowledge of sustainability, their attitudes to it, as well as the pedagogical approaches they use to educate their students. Crosling et al.'s study resulted in a list of pedagogical techniques that are used to conduct education on sustainability development. The most frequently used techniques they revealed include case study analyses, experiments, scenario development and analysis, organizing sustainability development days (at local, regional, and national levels), training sessions and awareness campaigns. Although this study was not dedicated to digital tools and not many applications of e-learning were mentioned, we believe that these conclusions are a great contribution both to climate change education in general and to digitalization of this education in particular. Most of the techniques presented in the study of Crosling et al. (2020) can be used either completely online or with blended learning. While being still effective for educational purposes, they would help diminish the carbon footprint by allowing students and teachers to stay at home instead of traveling to their place of study/work.

With a need to take a closer look at carbon footprints and traveling issues, as an important part of climate change awareness increase and education we could refer to the work of Nikula et al. (2023), which explores the internationalization of higher education and offers a few valuable observations. On the one hand, it turns out to be more emission-intensive to send teachers abroad to work in joint programs or other forms of transnational education than to employ local teaching staff. On the other hand, sending teachers abroad may have a smaller environmental cost than international travel for a large number of students. Although international transport was not the topic of our research, we believe that our work, together with the other studies discussed in this paper, may contribute to the formation of effective principles of environmental education and, moreover, of environmental behavior in general.

Finally, an interesting conclusion was drawn from Kelly et al.'s (2023) research about teaching and learning for sustainability science. These authors revealed a connection between people's willingness to take action to support green policies and their previous experiences with the consequences of climate change. People may feel separated from the effects of climate change because they either live far from places which, as they believe, are most affected by climate change, or because they think climate change is something that will happen in the future. While reducing physical distance to the consequences of climate change is hardly possible, it is important to reduce people's psychological distance and raise awareness about sustainability and climate change, among others, through learning at HEIs. In addition to that, Yu et al. (Yu et al., 2020) highlighted the necessity of not only raising awareness but also building and increasing students' motivation to undertake pro-environmental actions (e.g., turning lights off after use or recycling garbage). In support of both these ideas, Fang (2021), after correlating students' awareness with their pro-environmental behavior, states that students' higher awareness of climate problems leads them to be more willing to take pro-environmental actions.

Conclusions

This scoping review identifies a significant but limited number of papers (61) published over the past five years that specifically addressed the use of digital tools in climate education at higher-education institutions. A diverse array of digital tools is being utilized in climate education, with game-based environments, video communication, MOOCs, simulations, and video materials being some of the most prevalent. The use of digital tools in climate education is found to be beneficial for increasing student motivation, facilitating international collaboration, enhancing knowledge on environmental issues, and providing up-to-date information. Tools like serious games and simulations are particularly noted for their effectiveness in engaging students with complex environmental issues. Despite the advantages, there are challenges in integrating digital tools into climate education, including structural barriers (such as a lack of interdisciplinary team teaching), cultural barriers (such as biases about specific disciplines), and financial constraints. The research underscores the importance of active, experiential, collaborative, and challenge-based learning approaches in climate education.

References

Adipat, S., Laksana, K., Busayanon, K., Ausawasowan, A. and Adipat, B. (2021). Engaging Students in the Learning Process with Game-Based Learning: The Fundamental Concepts. *International Journal of Technology in Education*, *4*(3), 542–552. https://doi.org/10.46328/ijte.169

Adnan, S. and Xiao, J. (2023). A scoping review on the trends of digital anatomy education. *Clinical Anatomy*, *36*(3), 471–491. https://doi.org/10.1002/ca.23995

Alla, K.R. and Chen, S. Der. (2017). Strategies for Achieving and Maintaining Green ICT Campus for Malaysian Higher Education Institutes. *Advanced Science Letters*, *23*(5), 3967–3971. https://doi.org/10.1166/asl.2017.8244

Apichayakul, P., Pachanapan, P., Vongkunghae, A. and Tantanee, S. (2020). Isolated Energy Management Learning Platform through Smart and Green Building Design: A Case Study of USIS Building, Naresuan University. *2020 International Conference and Utility Exhibition on Energy, Environment and Climate Change (ICUE)*, 1–5. https://doi.org/10.1109/ICUE49301.2020.9307045

Arksey, H. and O'Malley, L. (2005). Scoping studies: Towards a methodological framework. *International Journal of Social Research Methodology: Theory and Practice*, *8*(1). https://doi.org/10.1080/1364557032000119616

Axelithioti, P., Fisher, R.S., Ferranti, E.J.S., Foss, H.J., Quinn, A.D., Dimopoulos, C., Meletiou-Mavrotheris, M., Boustras, G., Katsaros, E., Axelithioti, P., Fisher, R.S., Ferranti, E.J.S., Foss, H.J. and Quinn, A.D. (2023). What Are We Teaching Engineers about Climate Change? Presenting the MACC Evaluation of Climate Change Education. *Education Sciences 2023, Vol. 13, Page 153*, *13*(2), 153. https://doi.org/10.3390/EDUCSCI13020153

Baer, H.A. (2023). Grappling with Climate Change and the Internationalization of Higher Education: An Eco-Socialist Perspective. *Journal of Studies in International Education*, *27*(4), 638–653. https://doi.org/10.1177/10283153231172024

Baptista, F., Lourenço, P., Fitas da Cruz, V., Silva, L.L., Silva, J.R., Correia, M., Picuno, P., Dimitriou, E. and Papadakis, G. (2021). Which are the best practices for MSc programmes in sustainable agriculture? *Journal of Cleaner Production*, *303*, 126914. https://doi.org/10.1016/j.jclepro.2021.126914

Barteit, S., Sié, A., Yé, M., Depoux, A., Louis, V.R. and Sauerborn, R. (2019). Lessons learned on teaching a global audience with massive open online courses (MOOCs) on health impacts of climate change: a commentary. *Globalization and Health*, *15*(1), 52. https://doi.org/10.1186/s12992-019-0494-6

Bartlett, M., Larson, J. and Lee, S. (2022). Environmental Justice Pedagogies and Self-Efficacy for Climate Action. *Sustainability*, *14*(22), 15086. https://doi.org/10.3390/su142215086

Bowers, C. (2006). *Revitalizing the commons: Cultural and educational sites of resistance and affirmation*. Lexington Books.

Burns, H. (2009). *Education as Sustainability : an Action Research Study of the Burns Model of Sustainability Pedagogy*. https://doi.org/10.15760/etd.942

Capetola, T., Noy, S. and Patrick, R. (2022). Planetary health pedagogy: Preparing health promoters for 21st-century environmental challenges. *Health Promotion Journal of Australia, 33*(S1), 17–21. https://doi.org/10.1002/hpja.641

Cicha, K. and Rutecka, P. (2023, October 5). Digital Tools for Innovative Higher Education Teaching—A Scoping Review of Empirical Studies. *Information Systems Development, Organizational Aspects and Societal Trends*. https://doi.org/10.62036/ISD.2023.22

Cotton, D., Winter, J., Allison, J.A. and Mullee, R. (2023). Visual images of sustainability in higher education: the hidden curriculum of climate change on campus. *International Journal of Sustainability in Higher Education, 24*(7), 1576–1593. https://doi.org/10.1108/IJSHE-09-2022-0315

Crosling, G., Atherton, G., Shuib, M., Rahim, A.A., Azizan, S.N. and Nasir, M.I. M. (2020). *The Teaching of Sustainability in Higher Education: Improving Environmental Resilience in Malaysia* (pp. 17–38). https://doi.org/10.1108/S2055-364120200000022002

de Gaulmyn, C. and Dupre, K. (2019). Teaching sustainable design in architecture education: Critical review of Easy Approach for Sustainable and Environmental Design (EASED). *Frontiers of Architectural Research, 8*(2), 238–260. https://doi.org/10.1016/j.foar.2019.03.001

De Stefani, P. and Han, L. (2022). An Inter-University CBL Course and Its Reception by the Student Body: Reflections and Lessons Learned (in Times of COVID-19). *Frontiers in Education, 7*. https://doi.org/10.3389/feduc.2022.853699

Dichev, C. and Dicheva, D. (2017). Gamifying education: what is known, what is believed and what remains uncertain: a critical review. *International Journal of Educational Technology in Higher Education, 14*(1), 9. https://doi.org/10.1186/s41239-017-0042-5

Elder, S., Wittman, H. and Giang, A. (2023). Building sustainability research competencies through scaffolded pathways for undergraduate research experience. *Elem. Sci. Anth., 11*(1). https://doi.org/10.1525/elementa.2022.00091

Emblen-Perry, K. (2019). *Engaging and Empowering Business Management Students to Support the Mitigation of Climate Change Through Sustainability Auditing* (pp. 549–573). https://doi.org/10.1007/978-3-030-32898-6_30

Esmail, S. and Matthews-Roper, M. (2022). Lights, Camera, Reaction: Evaluating Extent of Transformative Learning and Emotional Engagement Through Viewer-Responses to Environmental Films. *Frontiers in Education, 7*. https://doi.org/10.3389/feduc.2022.836988

Falkenberg, L.J. and Joyce, P.W.S. (2023). Internationalisation at Home: Developing a Global Change Biology Course Curriculum to Enhance Sustainable Development. *Sustainability, 15*(9), 7509. https://doi.org/10.3390/su15097509

Fang, S.-C. (2021). The Pro-Environmental Behavior Patterns Of College Students Adapting To Climate Change. *Journal of Baltic Science Education, 20*(5), 700–715. https://doi.org/10.33225/jbse/21.20.700

Favier, T., Van Gorp, B., Cyvin, J.B. and Cyvin, J. (2021). Learning to teach climate change: students in teacher training and their progression in pedagogical content knowledge. *Journal of Geography in Higher Education, 45*(4), 594–620. https://doi.org/10.1080/03098265.2021.1900080

Fuertes-Camacho, M., Graell-Martín, M., Fuentes-Loss, M. and Balaguer-Fàbregas, M. (2019). Integrating Sustainability into Higher Education Curricula through the

Project Method, a Global Learning Strategy. *Sustainability, 11*(3), 767. https://doi.org/10.3390/su11030767

Greenwood, D.A. and Hougham, R.J. (2015). Mitigation and adaptation: Critical perspectives toward digital technologies in place-conscious environmental education. *Policy Futures in Education, 13*(1), 97–116. https://doi.org/10.1177/1478210314566732

Gregory, K.J. and Lewin, J. (2023). Big ideas in the geography curriculum: nature, awareness and need. *Journal of Geography in Higher Education, 47*(1), 9–28. https://doi.org/10.1080/03098265.2021.1980867

Jaleniauskiene, E. and Kasperiuniene, J. (2023). Infographics in higher education: A scoping review. *E-Learning and Digital Media, 20*(2), 191–206. https://doi.org/10.1177/20427530221107774

Jeong, J.S., González-Gómez, D., Conde-Núñez, M. C., Sánchez-Cepeda, J.S. and Yllana-Prieto, F. (2021). Improving climate change awareness of preservice teachers (Psts) through a university science learning environment. *Education Sciences, 11*(2). https://doi.org/10.3390/educsci11020078

Kelly, O., White, P., Butera, F., Illingworth, S., Martens, P., Huynen, M., Bailey, S., Schuitema, G. and Cowman, S. (2023). A transdisciplinary model for teaching and learning for sustainability science in a rapidly warming world. *Sustainability Science, 18*(6), 2707–2722. https://doi.org/10.1007/s11625-023-01407-z

Kousar, S., Afzal, M., Ahmed, F. and Bojnec, Š. (2022). Environmental Awareness and Air Quality: The Mediating Role of Environmental Protective Behaviors. *Sustainability, 14*(6), 3138. https://doi.org/10.3390/su14063138

Leal Filho, W., Aina, Y.A., Dinis, M.A.P., Purcell, W. and Nagy, G.J. (2023). Climate change: Why higher education matters? *Science of The Total Environment, 892*, 164819. https://doi.org/10.1016/J.SCITOTENV.2023.164819

Leichenko, R. and O'Brien, K. (2020). Teaching climate change in the Anthropocene: An integrative approach. *Anthropocene, 30*, 100241. https://doi.org/10.1016/j.ancene.2020.100241

Levac, D., Colquhoun, H. and O'Brien, K.K. (2010). Scoping studies: advancing the methodology. *Implementation Science, 5*(1), 69. https://doi.org/10.1186/1748-5908-5–69

Li, Y.Y. and Liu, S.C. (2022). Examining Taiwanese students' views on climate change and the teaching of climate change in the context of higher education. *Research in Science & Technological Education, 40*(4), 515–528. https://doi.org/10.1080/02635143.2020.1830268

Loeber, A., Van Mierlo, B., Grin, J. and Leeuwis, C. (2007). The practical value of theory: Conceptualising learning in the pursuit of a sustainable development. *In*: A.E.J. Wals (ed.). *Social Learning Towards a Sustainable World: Principles, Perspectives, and Praxis*. Brill | Wageningen Academic. https://doi.org/10.3920/978-90-8686-594-9

Maloney, K.M., Peace, A.L., Hansen, J., Hum, K.L., Nielsen, J.P., Pearson, K.F., Ramharrack-Maharaj, S., Schwarz, D.M., Papangelakis, E. and Eyles, C.H. (2023). Earth Science Education #7. GeoTrails: Accessible Online Tools for Outreach and Education. *Geoscience Canada, 50*(3), 73–84. https://doi.org/10.12789/geocanj.2023.50.198

Martek, I., Hosseini, M.R., Durdyev, S., Arashpour, M. and Edwards, D.J. (2022). Are university "living labs" able to deliver sustainable outcomes? A case-based appraisal of Deakin University, Australia. *International Journal of Sustainability in Higher Education, 23*(6), 1332–1348. https://doi.org/10.1108/IJSHE-06-2021-0245

Martí-Parreño, J., Seguí-Mas, D. and Seguí-Mas, E. (2016). Teachers' Attitude towards and Actual Use of Gamification. *Procedia - Social and Behavioral Sciences, 228,* 682–688. https://doi.org/10.1016/j.sbspro.2016.07.104

Membrillo-Hernández, J., Cuervo-Bejarano, W.J., Mejía-Manzano, L.A., Caratozzolo, P. and Vázquez-Villegas, P. (2023). Global Shared Learning Classroom Model: A Pedagogical Strategy for Sustainable Competencies Development in Higher Education. *International Journal of Engineering Pedagogy, 13*(1). https://doi.org/10.3991/ijep.v13i1.36181

Membrillo-Hernández, J., Cuervo-Bejarano, W.J. and Vázquez-Villegas, P. (2023). Digital Global Classroom, a Collaborative Online International Learning (COIL) Approach: An Innovative Pedagogical Strategy for Sustainable Competency Development and Dissemination of SDGs in Engineering Higher Education. *Lecture Notes in Networks and Systems, 633 LNNS,* 25–35. https://doi.org/10.1007/978-3-031-26876-2_3

Missimer, M. and Connell, T. (2012). Pedagogical Approaches and Design Aspects To Enable Leadership for Sustainable Development. *Sustainability: The Journal of Record, 5*(3), 172–181. https://doi.org/10.1089/SUS.2012.9961

Mohan, A., Mills, W. and Mohan, L. (2023). Multidisciplinary Approach to Teaching about Climate Change. *The Geography Teacher, 20*(3), 127–131. https://doi.org/10.1080/19338341.2023.2261463

Mulla, P. and Ratnayake, H. (2020). Innovative Transitioning To A Paperless Classroom. *INTED2020 Proceedings, 1.* https://doi.org/10.21125/inted.2020.0635

Nadeem, O. and Nawaz, M. (2023). Climate change and sustainable development perceptions of university students in Lahore, Pakistan. *International Research in Geographical and Environmental Education, 32*(3), 181–196. https://doi.org/10.1080/10382046.2022.2154973

Nikula, P.-T., Fusek, A. and van Gaalen, A. (2023). Internationalisation of Higher Education and Climate Change: A Cognitive Dissonance Perspective. *Journal of Studies in International Education, 27*(4), 586–602. https://doi.org/10.1177/10283153221145082

Ofori, B.Y., Ameade, E.P.K., Ohemeng, F., Musah, Y., Quartey, J.K. and Owusu, E.H. (2023). Climate change knowledge, attitude and perception of undergraduate students in Ghana. *PLOS Climate, 2*(6), e0000215. https://doi.org/10.1371/journal.pclm.0000215

Palmeiro-Silva, Y.K., Ferrada, M.T., Flores, J.R. and Cruz, I.S.S. (2021). Cambio climático y salud ambiental en carreras de salud de grado en Latinoamérica. *Revista de Saúde Pública, 55,* 17. https://doi.org/10.11606/s1518-8787.2021055002891

Parry, S. and Metzger, E. (2023). Barriers to learning for sustainability: a teacher perspective. *Sustainable Earth Reviews 2023 6: 1, 6*(1), 1–11. https://doi.org/10.1186/S42055-022-00050-3

Pavlova, E., Ulanova, K., Valeeva, N. and Zakirova, Y. (2020). Virtual Reality as a New Tool in Teaching English for Specific Purposes to Ecology Students. *Inted2020 Proceedings, 1.* https://doi.org/10.21125/inted.2020.1547

Seatter, C.S. and Ceulemans, K. (2017). Teaching Sustainability in Higher Education: Pedagogical Styles that Make a Difference. *Canadian Journal of Higher Education, 47*(2), 47–70. https://doi.org/10.47678/cjhe.v47i2.186284

Senevirathne, M., Amaratunga, D., Haigh, R., Kumer, D. and Kaklauskas, A. (2022). A common framework for MOOC curricular development in climate change education - Findings and adaptations under the BECK project for higher education institutions in

Europe and Asia. *Progress in Disaster Science, 14*, 100222. https://doi.org/10.1016/j.pdisas.2022.100222

Senevirathne, M., Priyankara, H.A.C., Amaratunga, D., Haigh, R., Weerasinghe, N., Nawaratne, C. and Kaklauskas, A. (2021). A capacity needs assessment to integrate MOOC-based climate change education with the higher education institutions in Europe and developing countries in Asia: findings of the focused group survey in PCHEI under the BECK project. *International Journal of Disaster Resilience in the Built Environment, 12*(5), 515–527. https://doi.org/10.1108/IJDRBE-07-2020-0074

Shields, R. and Lu, T. (2023). Uncertain futures: climate change and international student mobility in Europe. *Higher Education*, 1–18. https://doi.org/10.1007/S10734-023-01026-8/FIGURES/1

Södervik, I. and Vilppu, H. (2021). Case Processing in the Development of Expertise in Life Sciences-What Can Eye Movements Reveal? In *Applying Bio-Measurements Methodologies in Science Education Research* (pp. 169–183). Springer International Publishing. https://doi.org/10.1007/978-3-030-71535-9_9

Sormunen, M., Heikkilä, A., Salminen, L., Vauhkonen, A. and Saaranen, T. (2022). Learning Outcomes of Digital Learning Interventions in Higher Education. *CIN: Computers, Informatics, Nursing, 40*(3), 154–164. https://doi.org/10.1097/CIN.0000000000000797

Stein, S. (2023). Universities confronting climate change: beyond sustainable development and solutionism. *Higher Education*, 1–19. https://doi.org/10.1007/S10734-023-00999-W/METRICS

Stoeth, A.M. and Carter, K. (2023). Climate change summit: testing the impact of role playing games on crossing the knowledge to action gap. *Environmental Education Research, 29*(12), 1796–1813. https://doi.org/10.1080/13504622.2022.2129043

Straßer, P., Kühl, M. and Kühl, S.J. (2023). A hidden curriculum for environmental topics in medical education: Impact on environmental knowledge and awareness of the students. *GMS Journal for Medical Education, 40*(3). https://doi.org/https://dx.doi.org/10.3205/zma001609

Thompson, A.W., Marzec, R. and Burniske, G. (2022). Climate BufferNet. *Landscape Journal, 41*(1), 45–60. https://doi.org/10.3368/lj.41.1.45

Tomas, L., Evans, N. (Snowy), Doyle, T. and Skamp, K. (2019). Are first year students ready for a flipped classroom? A case for a flipped learning continuum. *International Journal of Educational Technology in Higher Education, 16*(1), 5. https://doi.org/10.1186/s41239-019-0135-4

Van Heuvelen, K.M., Daub, G.W., Hawkins, L.N., Johnson, A.R., Van Ryswyk, H. and Vosburg, D.A. (2020). How Do I Design a Chemical Reaction To Do Useful Work? Reinvigorating General Chemistry by Connecting Chemistry and Society. *Journal of Chemical Education, 97*(4), 925–933. https://doi.org/10.1021/acs.jchemed.9b00281

Vázquez-Vílchez, M., Garrido-Rosales, D., Pérez-Fernández, B. and Fernández-Oliveras, A. (2021). Using a Cooperative Educational Game to Promote Pro-Environmental Engagement in Future Teachers. *Education Sciences, 11*(11), 691. https://doi.org/10.3390/educsci11110691

Versteijlen, M. and Wals, A.E.J. (2023). Developing Design Principles for Sustainability-Oriented Blended Learning in Higher Education. In *Sustainability (Switzerland)* (Vol. 15, Issue 10). https://doi.org/10.3390/su15108150

Wade, A.A., Grant, A., Karasaki, S., Smoak, R., Cwiertny, D., Wilcox, A.C., Yung, L., Sleeper, K. and Anandhi, A. (2020). Developing leaders to tackle wicked problems at the nexus of food, energy, and water systems. *Elementa: Science of the Anthropocene*, *8*. https://doi.org/10.1525/elementa.407

Wan, Y., Ding, X. and Yu, H. (2023). Pre-service chemistry teachers' understanding of knowledge related to climate change. *Chemistry Education Research and Practice*, *24*(4), 1219–1228. https://doi.org/10.1039/D3RP00024A

Xabregas, Q.F. and Brasileiro, T.S.A. (2023). *Climate change in the extension curricularization in Higher Education Institutions (HEIs)*. Seven Editora. https://doi.org/10.56238/methofocusinterv1-075

Yeung, S.P.M. (1998). Environmental Consciousness among Students in Senior Secondary Schools: the case of Hong Kong. *Environmental Education Research*, *4*(3), 251–268. https://doi.org/10.1080/1350462980040302

Yu, T.-K., Lavallee, J.P., Di Giusto, B., Chang, I.-C. and Yu, T.-Y. (2020). Risk perception and response toward climate change for higher education students in Taiwan. *Environmental Science and Pollution Research*, *27*(20), 24749–24759. https://doi.org/10.1007/s11356-019-07450-7

CHAPTER 5

Fostering Climate Consciousness through Mathematics

Innovative STEM Activities for Pre-Service Teachers

İpek Saralar-Aras[1,*] and *Belma Türker Biber*[2]

Introduction

In the ever-evolving landscape of the 21st century, the intricate dance between climate change, education, and technology has emerged as a defining nexus for societal progress (Monroe et al., 2019). In an era marked by an urgent necessity for comprehensive climate literacy (Wynes and Nicholas, 2019), this chapter introduces a pioneering approach to address this pressing challenge. It does so by empowering pre-service teachers with tools to design STEM activities firmly grounded in climate and mathematical principles. Recognizing the intrinsic interplay between climate change and mathematical concepts, the chapter unveils a collection of carefully curated activities that seamlessly fuse

[1] Ministry of National Education, Ankara, Turkey.
[2] Aksaray University, Aksaray, Turkey.
Email: belmaturkerbiber@gmail.com
* Corresponding author: ipek.saralararas@gmail.com

mathematical problem-solving with ecological consciousness. These activities not only equip the educators of tomorrow with invaluable teaching resources, but also nurture a profound comprehension of the intricate bonds between numerical abstraction and environmental realities.

Integrating Climate Change Education in Middle School Mathematics

Climate change is a critical global issue that requires urgent action. Raising awareness about climate change, understanding critical situations, and addressing social impacts arenecessary to address this global problem. Education is an effective tool for raising climate awareness at an early age (Filho et al., 2023; Lee et al., 2013). Integrating climate change education across various grade levels and subjects, beginning with primary education, is methodized towards inculcating consciousness of the impacts on and consequences for the climate (Henderson et al., 2017; Ramazani and Saberi, 2023). Research shows that early exposure to climate concepts enhances student understanding and fosters a sense of responsibility toward environmental issues (Güngören et al., 2017; Savas et al., 2015). For this reason, many countries have included climate change in their curricula. This approach integrates climate change education into school lessons through various approaches ensuring that the problems posed to the climate are comprehensively addressed and understood (Henderson et al., 2017).

As an important stage in student cognitive development, secondary school education provides an opportunity to include climate consciousness as part of the curriculum. Integrating climate consciousness, interdisciplinary in nature, into various courses instills fundamental knowledge in students demonstrating an integrated educational approach that transcends traditional subject boundaries (Teksöz and Demirci, 2017). In this context, the integration of mathematical problem-solving with ecological consciousness not only enhances students' quantitative skills but also cultivates a holistic understanding of the real-world implications of climate change.

Since students spend most of their time in schools, lessons enriched with activities centred around climate change are valuable initiatives. There is an urgent need for teachers to design lessons that introduce climatic concepts and develop an awareness of the changing climate (Steffensen et al., 2021). Moreover, mathematics teachers must acquire knowledge about climate change and embed it in their pedagogy, given the clear correlation to mathematical principles (Barwell and Ernest, 2018). Barwell (2013) emphasizes that mathematics is a useful tool for understanding, describing, and predicting climate change. Mathematics provides effective models that predict climate evolution on a global scale (Barwell and Hauge, 2021). It is used to make sense of varied data obtained. In the simplest sense, mathematical figures, numbers, and graphs are used to represent data. It is therefore indispensable to employ mathematics to interpret this data meaningfully. Moreover, even though climate change is an interdisciplinary subject, it is a clear example of the use of mathematics in daily life.

Mathematics teachers and preservice mathematics teachers must improve themselves to develop consciousness of climate change in students, as one of the daily life connections of mathematics. Various approaches can be used to instil this consciousness in students, including the STEM approach, which relates mathematics to interdisciplinary learning. However, the mere existence of these approaches is insufficient if educators are not cognizant of their existence. It is essential that teachers are provided with training in these approaches. This chapter's emphasis on the use of STEM activities to foster climate consciousness aligns with research highlighting the efficacy of hands-on, experiential learning in middle school settings (Esen and Saralar-Aras, 2021; Saralar-Aras et al., 2022). Studies indicate that engaging students in collaborative learning environments with practical applications of mathematical concepts related to climate science, not only enhances academic performance but also stimulates a deeper interest in environmental issues (Crook, 1996), especially when also supported by technology (Crook, 2015). In the 21st-century landscape, the intersection of climate change, education, and technology has become pivotal for societal progress (Dere and Çinikaya, 2023). Through STEM activities, pre-service teachers and secondary school students can gain a solid understanding of climate change. In this context, this chapter contributes to the ongoing dialogue on effective pedagogical approaches that prepare pre-service teachers on the complex challenges posed by the climate crisis. We do this by reviewing studies on the integration of climate change education into middle school mathematics.

Geometry in Action: Illuminating Climate Change in Middle School Mathematics Education

This chapter explores the particular use of geometry in case studies of integrating climate change education within middle school mathematics. In this context, each of the STEM activities devised by pre-service teachers, which emphasize both climate change and geometry, constitutes a case. Geometry, with its emphasis on visual representation and spatial relationships, offers a unique lens through which to explore environmental concepts. Research suggests that visualizing abstract ideas enhances student comprehension and retention (Ainsworth, 2006; Ainsworth et al., 2011; Van Someren et al., 1998). Applied to climate change, geometric principles can illustrate phenomena like changes in landforms, the geometry of ecosystems, and spatial dimensions of environmental processes.

Middle school students often seek *relevance* in their learning experiences (Dickinson et al., 2011; Smid, 2023). Geometry provides a platform for real-world applications that enables students to connect mathematical abstractions to tangible environmental issues. By integrating geometric principles with climate change scenarios, the courses bridge the gap between theoretical knowledge and practical implications, thereby fostering a deeper understanding of the environmental challenges at hand.

Geometry inherently involves an interdisciplinary approach, drawing connections between mathematics and various other fields (Leahey, 1999; Saralar-Aras and Esen, 2022). By choosing geometry as the cornerstone for case studies, educators can seamlessly integrate climate science, mathematics, and technology, and thus reinforce the interconnectedness of these disciplines. This interdisciplinary perspective aligns with contemporary educational goals, encouraging students to view complex issues holistically.

Applying geometric principles to climate-related problem-solving cultivates essential skills such as critical thinking and analytical reasoning (Kuzle and Glasnović Gracin, 2019; Oldakowski and Johnson, 2018). Case studies that challenge students to analyze and solve real-world environmental issues using geometric concepts contribute to the development of a problem-solving mindset. This not only enhances their mathematical proficiency but also equips them with valuable skills for addressing complex challenges in the broader context of climate change.

Methodology

Research design

The methodology employed in this chapter is guided by a case study approach (Hollweck, 2015; Yin, 2014), designed to provide a comprehensive understanding of the integration of mathematics, climate change awareness, and pedagogical innovation within the context of pre-service teacher education. This approach combines six cases to capture the depth of participant experience within implemented STEM activities.

Participants

The study focuses on a diverse cohort of pre-service teachers enrolled in a teacher education program. The cohort comprised third-year preservice mathematics teachers, whose academic trajectory encompassed coursework in instructional principles and methodologies, including STEM education, technology integration, materials development, computer education, and information technologies. A purposive sampling strategy was employed to select participants with a range of mathematical backgrounds and varying levels of prior exposure to climate change concepts. The final sample included 6 participants, ensuring a representative cross-section of the target population working on geometry concepts among registered students at the time.

Data collection

The participants prepared STEM activities. Semi-structured interviews were conducted with pre-service teachers, before and after, who then engagemed with the created activities. The interviews aimed to acertain perceptions, attitudes, and understanding of the integration of mathematics and climate change concepts. Additionally, participants were encouraged to reflect on their experiences and pedagogical insights gained through the activities. STEM activities were conducted as a specific component within a 14-week course. This segment, spanning four weeks, was dedicated to STEM education, where in participants engaged in the development and implementation of STEM-related activities. The temporal interval between pre- and post-participation interviews facilitated an examination of the impact over the designated period.

STEM activity design and implementation

The core methodology of this chapter lies in the development and execution of innovative STEM activities, as in Zailskaite and colleagues (2021) that intertwine mathematical principles with climate change concepts. The process of designing these activities involved the following steps:

- Identification of Core Mathematical Concepts: The authors, who are subject matter experts, identified key mathematical areas suitable for integration with climate change concepts. These areas included but were not limited to geometry, algebra, statistics, and data analysis. For this chapter, we chose the ones related to geometry.
- Development of Activity Prototypes: Prototypical STEM activities were created by the participants, each linking a specific mathematical concept with a relevant aspect of climate change. These activities were designed to be engaging, hands-on, and adaptable for various grade levels.
- Piloting: The prototype activities were piloted with a smaller group of pre-service teachers (peers) to assess their feasibility, educational efficacy, and alignment with learning objectives.
- Refinement: Based on feedback from peers, the activities were refined and finalized for broader implementation.
- Feedback: The participants were interviewed on their experience of the process where they engaged in these activities during dedicated sessions.

The cases: STEM activities

The study analyzed the cases of six participants who designed STEM activities emphasizing both geometry and climate change, with each case referring to the STEM tasks prepared by the pre-service teachers.

These cases are described in detail in Table 1.

Data analysis

Interview transcripts were analyzed using thematic analysis to identify recurring themes and patterns in the participants' reflections and experiences. This allowed for a rich understanding of the qualitative aspects of their engagement with the STEM activities. Interview findings were supported with the representative examples from STEM activities.

Table 1: Preservice math teachers' Climate Change themed STEM activities

Title	Description
1. Climate Cartographers: Mapping Change with Geometry	In this hands-on STEM activity, middle school students will become climate cartographers, using their creativity and geometry skills to visually represent the effects of climate change on a local environment. Through the use of physical models and mapping techniques, students will gain a deeper understanding of geometric concepts while addressing the topic of climate change.
2. Climate Quest: Virtual Geometry Expedition	In this innovative STEM activity, middle school students will embark on a virtual geometry expedition to explore the effects of climate change on various environments. Using online tools and geometry concepts, students will navigate through interactive simulations to analyze the impact of climate change on different geographic features.
3. GeoTech Detectives: Mapping Climate Change Effects	In this technology-focused STEM activity, middle school students will become GeoTech Detectives, using digital tools and geometry to map and analyze the effects of climate change on their local environment. Through the integration of technology and mathematics, students will explore the real-world impact of climate change on geographical features.
4. Climate Crusaders: Geometry Expedition	In this STEM activity, middle school students will embark on a geometry expedition as climate crusaders to investigate and analyze the impact of climate change on their local environment. Through hands-on exploration and mathematical calculations, students will develop a deeper understanding of geometric concepts while raising awareness about climate change issues.
5. Geometry Detectives: Uncovering Climate Change Effects	In this STEM activity, middle school students will become geometry detectives as they explore the real-world effects of climate change on various geometric shapes and structures. By combining mathematics and environmental awareness, students will develop a deeper understanding of both subjects.
6. GeoClimate Explorers: Virtual Geometry and Environmental Change	In this technology-driven STEM activity, middle school students will become GeoClimate Explorers, using digital tools to virtually explore the effects of climate change on different environments. Through a combination of virtual simulations and geometry concepts, students will gain insights into the impact of climate change on various geographic features.

Ethical considerations

This research adhered to ethical guidelines, ensuring informed consent from all participants, confidentiality, and anonymity.

Findings

The findings of the study suggest profound possibilities in the integration of climate change awareness into mathematics within pre-service teacher education. In general, they illustrate the transformative role of innovative pedagogical approaches, as compared to traditional teaching methods, leading the way toward a more holistic mindset that intertwines mathematical principles with critical environmental consciousness. The participants continue to explore and develop these integrative methods in their future careers, committing to a process that enables knowledge to progress beyond the classroom. This exploration not only cultivates environmentally responsible students but also positions educators as stewards of societal change, imparting a sense of responsibility that extends far beyond mathematical concepts, shaping a generation attuned to the quantitative dimensions of climate change and its lasting implications for global sustainability. The following sections present the results thematically.

Climate change awareness and interdisciplinary thinking

Engagement with the STEM activities had a profound impact on the participants' climate change awareness and their ability to connect mathematical concepts with ecological issues. Qualitative data from interviews revealed several key findings:

Firstly, all six of the pre-service teachers expressed a heightened awareness of the real-world implications of climate change, emphasizing the significance of quantifiable data in understanding environmental challenges. All the activities aimed to foster a sense of responsibility and motivation among pre-service teachers to incorporate climate change themes into their future classrooms, thereby contributing to broader environmental education efforts. Moreover, almost all of the participants (5 out of 6) showcased an increased ability to view climate change through an interdisciplinary lens, appreciating the interplay between mathematics and ecological systems. They demonstrated an enhanced ability to communicate this interconnectedness to future students.

Participants' reflections on the integration of mathematics and climate change revealed nuanced insights. Some quotes are as follows: (P1) "I never thought geometry could be so connected to real-world issues like climate change. I prepared an activity on 'Climate Cartographers', and it changed my perspective on teaching mathematics," and (P4) "The STEM activities [Climate Crusaders: Geometry Expedition] allowed us to bridge the gap between abstract math and tangible problems. I see the relevance now, and so did my peers." (Refer to Table 1 for the details of the activities). With these statements, the pre-service teachers underlined the transformative impact of the integration, emphasizing a newfound appreciation for the practical application of mathematical concepts in addressing environmental challenges.

Pedagogical insights and adaptability to future classrooms

Participants provided valuable insights into pedagogical aspects of integrating mathematics and climate change in their future classrooms:

Almost all participating pre-service teachers (5 out of 6) reported that the hands-on nature of the STEM activities made complex mathematical concepts more accessible to students. Over half of the participating pre-service teachers (4 of them) recognized the adaptability of these activities across different grade levels, and the potential for customization to suit diverse learning styles and abilities. The overall experience highlighted the importance of fostering critical thinking skills, problem-solving abilities, and data literacy among students; vital aspectsof addressing climate challenges.

Further notable insights conveyed via interview include: (P2) "Designing and implementing these activities [Climate Quest] made me more aware of how students learn. It's not just about teaching math; it's about fostering critical thinking about real-world issues," and (P5) "Seeing the students' excitement during these sessions [Geometry Detectives] will make me realize the potential of hands-on, interdisciplinary learning, as in the teaching to peers." These statements emphasize potential pedagogical shifts toward student-centered, interdisciplinary approaches, and the broader educational impact possible beyond traditional mathematics instruction.

Long-term impact and sustainability

Over half of the pre-service teachers (4 of them) expressed a commitment to continue exploring innovative pedagogical approaches and the integration of mathematics and climate change in their future careers. As one participant passionately articulated, (P1) "I can't imagine going back to traditional teaching methods after this experience [Climate Cartographers]. Integrating mathematics with climate change isn't just a lesson; it's a mindset. I'm committed to using these innovative approaches in my teaching career to foster a generation of students who not only excel in math but also grasp the vital connection between numbers and our planet's sustainability." This sentiment reflects a dedication to transformative teaching practices that extend beyond the immediate classroom setting.

Another participant emphasized the broader societal impact, stating, (P2) "The impact of these activities [Climate Quest] goes beyond the classroom; it's about nurturing a sense of responsibility in our students. As future teachers, we carry the responsibility to mold environmentally conscious citizens. I see myself continuously exploring and refining these pedagogical methods to ensure a lasting impact on students' understanding of climate change through the lens of mathematics.". These perspectives consider teachers' roles not only in terms of academic growth but also in shaping responsible citizens who comprehend the quantitative aspects of climate change and its significant impacts on the long-term sustainability of our planet.

Impact on student engagement and understanding

Five participants out of six expressed a profound belief in the transformative ability of the STEM activities on student engagement and understanding.As one participant, (P3), noted, "Students who usually find math boring could actively participate and ask questions." This observation suggests the potential of STEM activities to invigorate the learning experience, challenging traditional perceptions of mathematics as tedious or abstract.

Another participant (P4), who prepared the 'Climate Crusaders: Geometry Expedition' activity, highlighted the broader significance, stating, "The integration will allow students to see the practical side of math. It isn't just about solving equations; it is about solving real problems that matter." These statements illuminate the transformative effect of integrative STEM activities on student perception, engagement, and the development of a broader understanding of the relevance of mathematics in addressing real-world issues.

Challenges and future directions

While the study yielded promising findings, all six of the pre-service teachers identified some challenges, including the need for additional resources and professional development to effectively implement these activities. Future research could explore strategies to address these challenges and further enhance the integration of mathematics and climate change education.

Participants acknowledged challenges in implementing STEM activities, as reflected in their statements: (P3) "Balancing the depth of mathematical content with climate change themes was challenging. It required careful planning to ensure both aspects were effectively communicated," and (P6) "Some students may initially struggle with the abstract nature of certain mathematical concepts, despite the real-life climate change relation. Adapting the activities based on their feedback will be crucial for successful implementation." These quotes highlight the dynamic nature of the implementation process, emphasizing the need for thoughtful adjustments to optimize the learning experience.

Discussion

The findings presented in this chapter demonstrate the transformative potential of integrating mathematics with climate change awareness within pre-service teacher education. This section delves into the implications of these findings, explores the broader significance of this innovative approach, and discusses its relevance in shaping future education strategies.

Strengthening climate literacy through mathematics

The improvement in participants' mathematical competence following engagement with the STEM activities underscores the value of interdisciplinary learning. Mathematics, often considered an abstract subject, gains new relevance when applied to real-world challenges (Dickinson et al., 2011). The findings suggest that integrating mathematical problem-solving with climate change themes not only enhances mathematical skills but also deepens climate literacy among pre-service teachers. As suggested by the academic literature on the subject (Barwell, 2013; Barwell and Hauge, 2021), this connection between abstract mathematical concepts and tangible ecological realities can serve as a model for effective environmental education in schools.

Fostering interdisciplinary thinking

The qualitative data reveals that pre-service teachers developed a greater appreciation of the interconnectedness of mathematics and climate change. This interdisciplinary perspective can be a catalyst for more holistic problem-solving abilities by students in the face of complex global issues (Filho et al., 2023; Lee et al., 2013). It empowers educators to bridge the gap between traditionally separate subjects and encourages students to view climate change through a multidisciplinary lens. This shift in thinking aligns with the evolving demands of the 21st-century workforce, where complex problem-solving often requires an interdisciplinary approach (Saralar-Aras and Esen, 2021; Saralar-Aras et al., 2022).

Pedagogical insights and student engagement

The pedagogical insights shared by participants highlight the potential of hands-on STEM activities to engage students effectively. By making mathematics relevant and applicable, these activities cater to diverse learning styles and abilities (Filho et al., 2023; Wynes and Nicholas, 2019). Moreover, they nurture critical thinking skills and data literacy, which are crucial not only for mathematics but also for informed citizenship in an era of data-driven decision-making. The experiences of pre-service teachers in designing and implementing these activities can serve as a source of inspiration and guidance for their future students.

Empowering future educators as agents of change

One of the most promising aspects of this study is the commitment expressed by pre-service teachers to continue integrating climate change themes into their future classrooms. This suggests that the impact of these activities extends beyond the immediate context of teacher education (Ateş et al., 2017; Teksöz and Öztekin, 2016). By empowering future educators with the tools and motivation to address climate change in their teaching, this approach has the potential to catalyze a broader cultural shift toward sustainability and environmental responsibility.

Addressing challenges and future directions

While the findings are encouraging, challenges were identified, including the need for additional resources and professional development. Addressing these challenges will be essential to scaling up the integration of mathematics and climate change education (Güngören et al., 2017; Teksöz and Öztekin, 2016). Future research can explore strategies for providing educators with the necessary support, materials, and training to effectively implement such activities in diverse educational settings. Additionally, exploring the long-term impact of these activities on student learning and environmental consciousness should be a priority for future studies.

Conclusion

In conclusion, the innovative integration of mathematics and climate change consciousness presented in this chapter offers a promising pathway to address the urgent need for comprehensive climate literacy. It empowers pre-service teachers to become agents of change, equipping them with the skills and perspectives needed to nurture environmentally conscious citizens. Thus, the study emphasizes the potential of mathematics as a powerful tool to address the complex challenges posed by climate change. By fostering interdisciplinary thinking, enhancing mathematical competence, and promoting pedagogical innovation, this approach holds the potential to shape a more sustainable and informed future.

References

Ainsworth, S., Prain, V. and Tytler, R. (2011). Drawing to learn in science. *Science, 333*(6046), 1096–1097. https://doi.org/10.1126/science.1204153

Ainsworth, S. (2006). DeFT: A conceptual framework for considering learning with multiple representations. *Learning and Instruction, 16*(3), 183-198. https://doi.org/10.1016/S0360-1315(99)00029-9

Ateş, D., Teksöz, G. and Ertepınar, H. (2017). Exploring the role of future perspective in predicting Turkish university students' beliefs about global climate change. *Discourse and Communication for Sustainable Education, 8*(1), 32–52. https://doi.org/10.1515/dcse-2017-0003

Barwell, R. (2013). The mathematical formatting of climate change: Critical mathematics education and post-normal science. *Research in Mathematics Education, 15*(1), 1–16. https://doi.org/https://doi.org/10.1080/14794802.2012.756633

Barwell, R. and Ernest, P. (2018). The philosophy of mathematics education today. In *Some thoughts on a mathematics education for environmental sustainability* (pp. 145–160). Springer.

Barwell, R. and Hauge, K.H. (2021). A Critical Mathematics Education for Climate Change: A Post-Normal Approach. In: *Applying Critical Mathematics Education* (pp. 166–184). Brill.

Crook, C. (2015). The 'digital native'in context: tensions associated with importing Web 2.0 practices into the school setting. In *Digital technologies in the lives of young people* (pp. 63–80). Routledge.

Crook, C. (1996). *Computers and the collaborative experience of learning.* Psychology Press.

Dere, İ. and Çinikaya, C. (2023). Tiflis bildirgesi ve BM 2030 sürdürülebilir kalkınma amaçlarının çevre eğitimi ve iklim değişikliği dersi öğretim programına yansımaları [Reflections of the Tbilisi declaration and UN 2030 sustainable development goals on the environmental education and climate change course curriculum]. *Ordu Üniversitesi Sosyal Bilimler Enstitüsü Sosyal Bilimler Araştırmaları Dergisi [Ordu University Social Sciences Institute Journal of Social Sciences Research], 13*(1), 1343–1366.

Dickinson, P., Hough, S., Searle, J. and Barmby, P. (2011). Evaluating the impact of a realistic mathematics education project in secondary schools. *Proceedings of the British Society for Research into Learning Mathematics, 31*(3), 47–52.

Esen, B. and Saralar-Aras, İ. (2021, October 22–24). Designing lesson plans for sustainable development focused education in the context of teaching polygons to middle school students. Paper presented at *the Second International Academician Studies Congress.* Karabük University, Turkey & Online, ASC 2021.

Filho, W.L., Balasubramanian, M., Zúñiga, R.A.A. and Sierra, J.B. (2023). The effects of climate change on children's education attainment. *Sustainability, 15*(7), 6320. https://doi.org/10.3390/su15076320

Güngören, S., Teksöz, G. and Şahin, E. (2017). Turkish middle school students level of attitude on the vital threat of their future: climate change. Presented at the *IX. International Congress of Educational Research.*

Henderson, J.V., Storeygard, A. and Deichmann, U. (2017). Has climate change driven urbanization in Africa?. *Journal of Development Economics, 124,* 60–82. https://doi.org/10.1016/j.jdeveco.2016.09.001

Hollweck, T. (2015). Review of Robert K. Yin. (2014). Case study research design and methods. *Canadian Journal of Program Evaluation, 30*(1), 108–110. https://doi.org/10.3138/cjpe.30.1.108

Kuzle, A. and Glasnović Gracin, D. (2019, July). Classroom social climate in the context of middle school geometry. In *Proceedings of the 43rd Conference of the International Group for the Psychology of Mathematics Education* (Vol. 2, pp. 511–518).

Leahey, L.K. (1999). *An interdisciplinary approach to integrated curriculum.* (Master's dissertation). Rowan University Theses and Dissertations, 1837. https://rdw.rowan.edu/etd/1837

Lee, T.M., Markowitz, E.M., Howe, P.D., Ko, C.Y. and Leiserowitz, A.A. (2015). Predictors of public climate change awareness and risk perception around the world. *Nature Climate Change, 5*(11), 1014–1020. https://doi.org/10.1038/nclimate2728

Monroe, M.C., Plate, R.R., Oxarart, A., Bowers, A. and Chaves, W.A. (2019). Identifying effective climate change education strategies: A systematic review of the research.

Environmental Education Research, 25(6), 791–812. https://doi.org/10.1080/13504622.2017.1360842

Oldakowski, R. and Johnson, A. (2018). Combining geography, math, and science to teach climate change and sea level rise. *Journal of Geography, 117*(1), 17–28. https://doi.org/10.1080/00221341.2017.1336249

Ramazani, S.H.R. and Saberi, M.H. (2023). Journal of Drought and Climate change Research (JDCR). *Journal of Drought and Climate Change Research (JDCR), 1*(2), 41–52. https://doi.org/10.22077/JDCR.2023.5977.1007

Saralar-Aras, İ., Esen, B. and Akdur, T.E. (2022). *Matematik ve geometri eğitiminde STEM çalışmaları rehberi [A guidebook on STEM studies in mathematics and geometry].* Ministry of National Education DGIET Publications. Ankara.

Saralar-Aras, İ. and Esen, B. (2021). Geometri eğitiminde STEM çalışmaları [STEM practices in geometry education]. pp. 207–232. *In*: İ. Saralar-Aras (ed.). *Okul öncesinden ortaöğretime farklı disiplinlerde STEM eğitimi uygulamaları [STEM education practices from kindergarten to secondary school].* Ankara, Türkiye: Millî Eğitim Bakanlığı D.S.İ./ Yenilik ve Eğitim Teknolojileri Genel Müdürlüğü.

Savas, G., Teksöz, G. and Şahin, E. (2015). Turkish middle school students level of knowledge on the vital threat of their future climate change. Presented at *the ESERA 2015*, Helsinki, Finland.

Steffensen, L., Herheim, R., Rangnes, T.E., Andersson, A. and Barwell, R. (2021). Applying critical mathematics education. pp. 185–209. *In*: *the Mathematical Formatting of How Climate Change Is Perceived: Teachers' Reflection and Practice*. Brill. https://doi.org/https://doi.org/10.1163/9789004465800_009

Smid, H.J. (2023). Realistic mathematics education. pp. 247–261. In *Theory and practice: A history of two centuries of Dutch mathematics education*. Springer International Publishing.

Teksöz, G. and Demirci, S. (2017). About the context of climate change education for sustainability: In the words of university students. Presented at the *ESERA 2017 Conference.*

Teksöz, G. and Öztekin, C. (2016). An attempt for developing determinants of low carbon behavior a clue for climate change adaptation and mitigation. Presented at *the 14th International Conference Sustainable Development, Culture, Education.*

Van Someren, M.W., Reimann, P., Boshuizen, H. and de Jong, T. (1998). *Learning with multiple representations: Advances in learning and instruction series.* Elsevier Science.

Wynes, S. and Nicholas, K. (2019). Climate science curricula in Canadian secondary schools focus on human warming, not scientific consensus, impacts or solutions. *Plos One, 14*(7), e0218305. https://doi.org/10.1371/journal.pone.0218305

Yin, R.K. (2014). Case study research design and methods (5th ed.). Sage.

Zailskaite, L., Burbaite, R. and Kutas, R. (2021). *STEAM framework for teacher training.* EDUSIMSTEAM: Erasmus+ KA3 Forward Looking Cooperation Project Document.

CHAPTER 6

Information and Communication Technologies in Climate Change Education

Octav-Sorin Candel, Oara Prundeanu, Nicoleta Laura Popa* and *Ștefan Boncu*

Introduction

Climate change represents a dramatic threat to humanity and, as such, in recent years has received a great amount of attention from politicians, policymakers, scientists and educators. Many top-down initiatives discuss resource conservation and try to create an ideal legislative environment for intervention. At the same time, bottom-up engagement is needed to mitigate the effects of climate change and foster the adoption of new practices of sustainable development (Anderson, 2012; Fernandez Galeote et al., 2021). In this context, climate change education (CCE) can become a primary means to promote climate-responsive actions (Cantell et al., 2019). However, despite strong institutional and individual initiatives in implementing CCE, there are important limitations that must be addressed (Leal Filho et al., 2021; Rousell and Cutter-Mackenzie-Knowles, 2020). Past reviews have

Alexandru Ioan Cuza University of Iași, Romania.
Emails: oara.prundeanu@uaic.ro; nicoleta.laura.popa@uaic.ro; boncu@uaic.ro
* Corresponding author: octav.candel@uaic.ro

shown that a deficiency in teacher knowledge of climate change, lack of inclusion of climate change in teaching programs and participatory, interdisciplinary, creative, and affect-driven approaches have hurt the propagation and effectiveness of CCE (Leal Filho et al., 2021; Cantell et al., 2019; Rousell and Cutter-Mackenzie-Knowles, 2020). Moreover, these issues appear at many levels, significantly in primary, secondary, and university education. Several ideas have been proposed to address these problems. To be effective, CCE must encompass personally relevant and meaningful information and active and engaging teaching methods (Monroe et al., 2019). As such, there is a pressing need for strong theoretical models of CCE, as well as for innovative ways of implementing content in both formal and informal contexts.

Most initiatives trying to improve CCE have been implemented in STEM education (Rousell and Cutter-Mackenzie-Knowles, 2020). While producing scientific knowledge and finding new technologies to mitigate climate change is essential (Cantell et al., 2019), educators can also use new technological advancements as tools for delivering information, promoting values, and motivating action. In recent years, Information and Communication Technologies (ICT) have emerged as a powerful modality for achieving CCE (Rousell and Cutter-Mackenzie-Knowles, 2020). However, few studies have tried to assess the extent to which they may be used and their potential impact, with an empirical theoretical framework. With the aim of covering both theoretical insights and practical implementation, this chapter aims to evaluate how well-suited is the use of ICT in this field. For this, we will draw on the lens of the Bicycle model on climate change education (Cantell et al., 2019), one of the more complex models aiming to describe and improve CCE.

The United Nations Framework Convention on Climate Change (UNFCCC) considers education, training and public awareness as important processes to deal with climate change. Moreover, education is seen as both an ethical as well as cost-effective approach when referring to climate-related issues (Mochizuki and Bryan, 2015). An advantage of CCE is that it can be approached in multiple ways; either top-down (short-term and long-term institutional and policy changes) or bottom-up (individuals empowered to design their own climate change projects). However, any such initiative requires specific objectives to create effective educational practices. For this reason, Tolppanen and colleagues (2017) created the *Bicycle Model*; a holistic approach that imagines components of climate change education similarly to the components of a bicycle. To be successful, all components are needed and important. In addition, the model takes into account not only information provided by the natural sciences (critical to dealing with climate change), but also from social sciences, health sciences and humanities (Cantell et al., 2019).

According to Cantell et al., (2019), the most essential component of CCE is knowledge and thinking skills (the wheels of the bicycle). Similarly, Anderson (2012) considers that critical thinking, problem solving, scientific literacy, education on sustainable lifestyles and consumption, disaster risk preparedness and green technical and vocational training to be integral components of CCE. However, to effectively respond to climate-related issues, knowledge alone is not sufficient. Individuals need to develop values and a worldview that is congruent with CCE, through empathy and a desire to help (the frame of the bicycle). Once these components are in place and the individuals are sufficiently motivated (the saddle of the bicycle), they can enact pro-environmental measures to mitigate climate change (the chains and pedals of the bicycle). Still, there exist operational barriers (the brakes), such as structural, psychological and socio-cultural problems, that can hamper environmental responsibility. To overcome such issues, the development of positive emotions (the lamp) and future orientation (the handlebar) are needed.

Although rooted in research, the bicyle model does not elaborate on ethods educators may employ to develop its components. Moreover, Rousell and Cutter-Mackenzie-Knowles (2020, p. 191) consider that "the participatory, interdisciplinary, creative, and affect-driven approaches to climate change education [...] have been largely missing from the literature." We propose that a fruitful approach to the subject is the utilization of Information and Communication Technologies. Other academic reviews have discussed the importance of using ICT (Monroe et al., 2019; Wu and Lee, 2015) and past studies have shown that it is effective in increasing climate change engagement, delivering pro-environmental information, and fostering pro-environmental attitudes and behaviours (Boncu et al., 2022; Fernandez Galeote et al., 2021). Moreover, ICT approaches can be used both in classrooms and outside them, and as a form of self-directed informal learning (Boncu et al., 2023), thus being beneficial to all types of learners, regardless of age and educational background. Given that lifelong learning, multiple levels of education (primary, secondary, tertiary and adult education), and a diversity of modes of delivery (formal, non-formal, informal) are crucial for effective CCE (Mochizuki and Bryan, 2015; Monroe et al., 2019), this study aims to discuss the use of ICT applied in regards to all aspects of the bicycle model.

Wheels: Knowledge and Thinking Skills

Knowledge is essential for effective CCE, but acquiring knowledge alone is not sufficient (Cantell et al., 2019). The accumulation of information must facilitate a deeper understanding of climate change, a process possible only through the use of critical thinking faculties. Still, most CCE interventions are primarily focused on the increase of subject knowledge (Monroe et al., 2019).

Nonetheless, ICT use can be an effective way to transmit the proper information about climate change both in formal educational environments, and informal ones. In recent years, various papers have described how curriculums can be improved using technology in STEM-related fields. Furthermore, these programs varied in terms of ease of implementation or use. Karpudewan and Mohd Ali Khan (2017) applied experiential learning to improve student knowledge about the environment. Technology, in the form of video materials and online calculators of carbon footprints, was used to augment real-life field trips and role-play. Other studies have used immersive virtual reality to *organize* virtual field trips to Greenland, in an exercise designed to increase declarative knowledge (knowing facts and information about climate change) self-efficacy and the intention to pursue an education and a career in natural sciences (STEM intentions) among students (Peterson et al., 2020). Finally, the *Heat-Cool Initiative* proposes the use of thermal imaging cameras to develop knowledge about thermodynamics and the effects of climate change on temperature among primary and secondary school students (Kumar et al., 2023). More complex systems can be also useful. For example, information from NASA satellites was used in a series of tutorials about the use of remote sensing in observing climate change (Cox et al., 2014). Similarly, in the *Arctic Climate Connections Curriculum*, students use data about arctic temperatures and images to gain better knowledge about climate change in the Arctics (Gold et al., 2015).

While these technologies are primarily used in classrooms, other forms of ICT are more flexible and can lead to promising results in more informal environments. *Serious* computer games and online applications are noteworthy methods that can contribute to gaining information and developing knowledge about climate change (Boncu et al., 2022; Wu Knowledge is essential for effective Lee, 2015). Gamification is a viable tool for CCE as it can be considered a life-long method of education, given that individuals of all ages have access to computer games and applications.

However, as previously mentioned, knowledge gain alone is not sufficient. Studies show that some CCE courses lead to significant increases in scientific knowledge, but not in knowledge of mitigative actions (Tolpponen et al., 2022). Also, climate change being a contentious political issue, political polarization, wide-spread climate denial and the development *post-truth* mentalities can impact the possibility of action (Rousell and Cutter-Mackenzie-Knowles, 2020; Sarewitz, 2011). Thus, CCE must also promote critical thinking skills that can help students and non-students alike navigate today's political and media environments. Unfortunately, the use of ICT for the development of critical thinking within CCE seems rather scarce. Several e-learning courses have been developed to teach critical thinking in relation to global warming and environmental change (Chusni et al., 2021; Suwatra et al., 2018). There have also been some programs on how to navigate current political environments. For example, the online course *Freirean Communicative Educational Situations for Climate Change Education*, which was offered to teachers, aimed to improve their critical thinking skills and enforce the idea that climate change is a political issue as much as it is a scientific one (Mejia-Caceres et al., 2023).

Frame: Identity, Values, and Worldview

Individuals who do not have environmental values and environmental identities cannot assimilate information about the negative aspects of climate change and will not be compelled to act towards its mitigation. In other words, personal worldviews must coincide with pro-environmental behaviours. Unless this is so, personal values, identity and worldviews will constitute blockages preventing individuals from perceiving the true cost of climate change.

Indeed, research has shown that an individual-centric worldview can lead to lower levels of acceptance of anthropogenic global warming (Stevenson et al., 2014). In addition, possessing a hierarchical and individualistic worldview seems to be related to the belief that climate change mitigation is unnecessary (Hornsey, 2021). However, the former study also shows that the worldviews of adolescents are more flexible and can be changed, as compared to those of adults (Stevenson et al., 2014). Thus, a concerned body of young people, engaged in CCE, may well lead to worldviews favourable toward climate change intervention. Unfortunately, there is almost no proof of how ICT can contribute to this process. Still, Cook (2019, p. 25) considers that "technology will undoubtedly give students an unprecedented, multidimensional space of options, opportunities and even realities". Social media and various apps can be used to engage the youth in activities regarding climate change (Rousell and Cutter-Mackenzie-Knowles, 2020; Rousell et al., 2023). In addition, offering information and developing critical thinking can increase acceptance of scientific truths and, in-turn, create favourable worldviews, however these actions must be carried out in a culturally sensitive manner (Lewandowsky, 2021).

Studies have demonstrated how identity and values can determine the effectiveness of climate change information campaigns. Bolderdijk and colleagues (2013), for example, presented a movie clip portraying the negative environmental consequences of using bottled water. However, the images inspired an intention to act, and acceptance of environmental policies related to plastic waste management, only in participants who already held biospheric values (i.e., those individuals with a well-defined concern for the environment).

Fortunately, via identification, ICT offers the possibility of choosing virtual identities different from those in everyday life, including some climate-related ones, and through them, facilitates changes in values and attitudes (Yee et al, 2009). The individuals may consider that their pro-environmental virtual identity is attractive and noble, thus they would want to act in ways that are more similar to their virtual character even in real life. For example, Fernández Galeote and colleagues (2022), examined roles assumed by individuals in 80 video game scenarios related to climate change. Six distinct types of identities were documented, each differing in terms of inclination, motivation, optimism and outlook towards climate change intervention. Thus, these alternative identities can empower individuals and provide the catalyst to shape their identities towards climate empathy.

Chains and Pedals: Action to Curb Climate Change

The body of knowledge individuals possess, and their values and identities, must be channeled into action for progress to be made towards protecting and preserving the environment. Through educational, often only the transmission of specialized knowledge is considered, with relatively few activities designed to empower individuals to apply their knowledge. Environmental knowledge has no significant direct effect on pro-environmental behaviour (PEB), but plays an important role in the formation of positive environmental attitudes (Liu et al., 2020). Thus, for the bicycle to start from a standstill, effort is required from the user to push the pedals. However, if a certain PEB is perceived as needing too great an effort, individuals tend not to perform it (Dreijerink et al., 2022). The concept of effort can be described from the perspective of the *actual* effort performed by an individual (i.e., objective effort) or from the perspective of effort subjectively *perceived* by the individual (Steele, 2020). Often, depending on the perceived effort, individuals choose to act based on whether or not the effort offers sufficient value and/or reward (Inzlicht et al., 2018). To address the issue ofindividual perceived effort, group-based environmental education interventions are promising tools to increase positive efficacy beliefs among younger generations.

Several mobile apps can facilitate the implementation of pro-environmental actions (e.g., Greencoin; Radziszewski et al., 2021). Eco-apps allow individuals to compare their performance with others, learn new behaviours and strategies for the implementation of PEB, and promote the development of individual pro-environmental routines (Ochmann and Lehrer, 2023). Educational mobile tools should foster habit formation (i.e., new behaviour becomes a habit through daily repetition) to ensure long-term behavioural change and maintenance (Judah et al., 2013; Ro et al., 2017). Also, the use of mobile applications can lead to long-term changes in the attitudes and behaviours of users. For example, Ro and colleagues (2017) tested the effectiveness of the *Cool Choices* game (i.e., game-based sustainability intervention in which players compete in teams to gain points for their sustainable actions). Results showed that users reduced household electricity consumption, and that the game led to the adoption of PEB, with effects lasting up to six months. Real-world action games for climate change education (e.g., GREENIFY; Lee et al., 2013) facilitate action-oriented learning and stimulate informed action. The GREENIFY system allows users to build daily missions related to actions they can take to reduce their impact on the climate, receive information about environmental problems, build action groups that share common visions and values (i.e., positive peer pressure) and create group-challenge experiences (e.g., setting goals, challenges) (Lee et al., 2013).

Immersive virtual reality (IVR) can be used as an alternative educational approach to improve waste management in classrooms (Stenberdt and Makransky, 2023). IVR can promote behavioural strategies related to energy-saving by allowing users to perceive, through a sensory experience, the impact of specific actions. As a result, individuals tend to apply at home the sustainable actions learned in IVR (Kleinlogel et al., 2023). In conclusion, students who use IVR technology in the classroom can experience climate change scenarios but also can experiment with the positive impact of personal choices by transferring their knowledge into real-world actions (Plechatá et al., 2022).

Stimulating Motivation and Participation

For the authors of the bicycle model (Cantell et al., 2019), motivation and participation are embodied by the bicycle seat. The saddle influences motivation through comfort and ease of use. Cantell and her colleagues (2019) view motivation and participation as key targets of CCE as potential stakeholders, young people in particular, do not see themselves being affected by climate change and, perhaps worse, do not consider themselves as agents of change.

Environmental motivation can be extremely weak, since, unlike other motivations, we do not inherit it through ancestory (Brick et al., 2021). In other words, concern for the state of the natural environment is hard to explain in evolutionary terms, and, as such, environmentalist beliefs, values, motivations and identities are difficult to generate and maintain.

There is a growing recognition of the role of intrinsic motivation as a catalyst for the acquisition of climate change knowledge, and forindividual behaviours towards mitigating its effects. Monroe and colleagues (2019), after reviewing a considerable number of CCE efforts, distinguished two main strategies that seem to ensure the success of such interventions: making climate change information personally relevant and meaningful for learners and using active and engaging teaching methods.

In terms of ICT interventions, remote sensing (RS) can be an effective and exciting method that not only enriches the knowledge of students, but also challenges and engages them. In their study, Asimakopoulou and colleagues (2021) used satellite *Remote Sensing for Earth Observation*. Their data came from teachers they interviewed after they worked with this an innovative method. They found "there is a high interest in how satellites depict environmental phenomena and that EO is an efficient vehicle for promoting climate change education in schools because it illustrates climate change impacts most effectively" (Asimakopoulou et al., 2021 p. 2). The teachers in their study had no prior knowledge of RS, but, at the conclusion of the exercise, "they recognized the potential of Earth-Observation as a very effective approach for attracting students' interest and engaging them in active learning" (Asimakopoulou et al., 2021, p. 12).

ICT can be used to both develop motivation in young people to participate in CCE as well as to identify existing sources of motivation. Ross and colleagues (2021) built a model of motivation for CCE called the *Holistic Agentic Climate-Change Engagement Model* (h-ACE). Participants in this study, students aged 13 to 15 from two schools in Wales, generated IDNs (interactive digital narratives) about climate change using *Twine* storytelling software. The qualitative analysis of these student-created interactive digital narratives provides valuable information about how teenagers participate in CCE sessions. One of the important conclusions of the study is that engagement was related to the participants' views on their capacity to produce change on individual, local and governmental levels.

Brakes: Operational Barriers

To achieve change in the daily behaviours of individuals, there is a need to analyse the diversity of factors representing operational barriers in implementation of pro-environmental behaviours (PEB). Perceived barriers include lack of time, money, low levels of self-efficacy and awareness, and counterproductive social norms in the absence of community facilities (Pruneau et al., 2006). In the analysis of the factors that influence involvement in PEB, Stern (2000) distinguishes between contextual factors (e.g., government regulations, monetary incentives and costs, social, economic, and political context), attitudinal factors (e.g., norms, beliefs, and values), personal capabilities (e.g., knowledge, skills, money, time, social status and power, sociodemographic variables) and individual habits or routine. To overcome these limits, behavioural catalysts should be identified, such as changing social norms through education and institutional support (Quimby and Angelique, 2011). For example, the *low-cost hypothesis* highlights the fact that the individual's behaviour depends on how they perceive associated costs, and that this influence on behaviour can change with an associated change in cost perception (Huang et al., 2020). For example, residents of a community are willing to participate in recycling activities as long as the perceived costs are low and most residents in the neighbourhood participate (Keuschnigg and Kratz, 2018). Therefore, when perceived costs are low, individuals tend to behave in accordance with the social norm in order to preserve their status within the group to which they belong.

Further, we can foster environmental awareness and perceived self-efficacy in educational contexts by using the *Internet of Things* approach (i.e., IoT systems). These systems are economically and ecologically efficient and can promote sustainable development. For example, Tabuenca and colleagues (2020) tested the effectiveness of *Smart IoT Planters* in university campuses throughout a semester by using multidisciplinary teams. Starting with the idea that the presence of plants in the classroom increases the well-being of students and makes them develop more positive attitudes towards plants, the researchers proposed a holistic architecture to implement IoT systems in plants. These systems contain sensors for plant monitoring, automatic irrigation, and artificial light (Tabuenca et al., 2020). In addition, they used the project-based-learning framework, aiming to develop real-world specific knowledge in a multi-disciplinary way. The teachers explained to the students the main objectives related to sustainable development (e.g., implementing urgent actions to combat climate change, ensuring access to modern, sustainable, and affordable energy methods, and ensuring sustainable consumption). The students built and implemented various sensors and used an IoT cloud platform to monitor data from planters. Results obtained by Tabuenca et al., (2020) indicated that students reported higher levels of environmental awareness across the campus. Finally, this study highlights the need to use technological solutions to increase the understanding of ecological problems, and to focus on specific solutions that students can implement to promote environmental awareness in university campuses.

Environmental awareness is often constrained by multiple cognitive and emotional limitations (Kollmuss and Agyeman, 2002). For example, perceived temporal distance (how much time one perceives between their current time and a future event) is a barrier highlighted in research analysing individuals' perceptions of climate change, because disastrous environmentalconsequences seem so temporally distant (Soliman et al., 2018). Furthermore, climate change is not immediately tangible, and ecological destruction is sometimes slow and gradual, which may lead some to believe that we are not currently facing significant environmental problems. To overcome these cognitive barriers, it is necessary for individuals to be able to anticipate the potential long-term consequences of their behaviours. In general, when people perform a behaviour, they want to immediately notice its effects, being driven by the need to obtain instant gratification. Through mobile applications, students could receive certain rewards as a result of achieving some PEBs, but not immediately. Rather, the reward would be granted after they achieved a certain target that would require a prolonged effort. Thus, the ability to delay gratification enabled students to perceive a greater gain from the investment made (Gschwandtner et al., 2022). Moreover, to overcome the temporal distance barrier, we can use various experimental approaches to change the subjective proximity of certain climate change effects (Gifford, 2011; Soliman et al., 2018).

In conclusion, to overcome the obstacles perceived by the individual in order to implement PEB, a holistic approach is necessary, starting from micro actions (e.g., opportunity-enhancing practices for students), to macro actions (e.g., implementing initiatives that contribute to the emergence of a pro-environmental culture in university or other educational institutions) (Akhtar et al., 2022). Educational institutions could become environmental hubs that promote sustainable visions related to the environment, offering students a conducive framework for PEB implementation. Also, university campuses should consider technological and energy optimization measures (e.g., monitoring energy use or implementation of sensor networks) to promote efficient energy consumption and to provide an example of good practices to students (Kolokotsa et al., 2016). CCE should focus on removing structural barriers wherever possible, by using a combination of psychological and technological interventions.

Lamp: Hope and other Emotions

As argued in various studies (Kerret et al., 2016; Cantell et al., 2019; Ojala, 2023a), emotions and emotional awareness are not only relevant but critical for successful CCE, as they may drive both general and emotional well-being given a positivity ratio is achieved (i.e., individuals experience more positive than negative emotions as a result of engagement in CCE programs and activities). Both negative and positive emotional states have been studied in relation to climate change and related educational endeavours, for various age groups and in different cultural contexts. A recent survey based on a large sample of children and young people, in ten countries across the world, reported prevalent and concerning incidences of negative emotions associated with climate distress and anxiety, such as sadness, fear, anger, powerlessness, helplessness, guilt, shame, despair, pain and grief (Hickman et al., 2021). On the other hand, teachers taking on the responsibility of CCE also experience negative emotions and face difficulties in framing the climate future in a hopeful manner, primarily due to self-imposed emotional constraint and repression (Beasy et al., 2023).

Although the control of negative emotions in CCE is constantly addressed in recent research (e.g., Lawrance et al., 2022), an increasing number of studies equally stress the importance of hope and hopefulness as balancing factors in supporting climate engagement (Geiger et al., 2023; Cantell et al., 2019; Kerret et al., 2016). Hope is a multifaceted state with cognitive, motivational and emotional components that alert people to the prospect of positive future results (Peterson and Seligman, 2004). Hope is often understood in terms of its cognitive and motivational components, which tend to follow hope's emotional experiences (Geiger et al., 2019). Also, the emotional components of hope indicate an anticipated condition that is frequently regarded as pleasurable (Geiger et al., 2023). Beyond different emphases on definitions and measures, researchers have reached a consensus on hope as a future-oriented state that leads people to foresee positive futures (Kantenbacher et al., 2022). Based on this line of reasoning, climate activists, educators and researchers explore the relationship between hope, positive emotions and pro-environmental behaviour or climate action. Kerret et al. (2016) showed the mediation effect of hope, moderated by self-control skills, between participation in green school activities and pro-environmental behaviour, as well as a positive environmental ratio among high school students. Similarly, Finnegan (2023) found a strong positive association between hope and motivation towards action among secondary school students. Also, teachers whose practices included both acceptance of negative emotions and ensuring positive outlooks were good predictors of inspiring studen environmental optimism. Teachers value hope as a powerful positive force in CCE, but they are also aware of the pitfall of envisioning over-optimistic futures (Ojala, 2022). A recent comprehensive meta-analysis conducted by Geiger et al. (2023) suggests that fostering hope may be over-evaluated as a predictor of climate engagement and action, but it is certainly associated positively with taking responsibility for climate action at an individual level.

Digital applications addressing sustainable futures gain popularity as powerful and appealing educational tools, and have the potential to foster hope or, at least, overcome hopelessness among individuals of all ages. For example, *Last Island* is a serious, collaborative-competitive game created to study potential sustainable futures, which tries to teach a non-expert community of players about sustainability and how an isolated society may transition to alternative futures (Taghikhah et al., 2019). It replicates variables critical to a society's sustainability, such as human population, economic production, and environmental state, and engages players in collaboratively finding and testing viable scenarios for transitioning to a sustainable future. *EnerCities* is a 3D game created by *Qeam* with support from the University of Twente and partners, designed to be implemented in European schools (Knol and DeVries, 2011) that empowers players to create a virtual sustainable city in a realistic scenario, with the main purpose of informing about renewable and non-renewable sources of energy. A mixed study conducted by Janakiraman et al. (2021) indicates that students playing the game developed environmentally friendly attitudes and behaviours, especially in terms of energy savings, but also reflected on their overall perspective about the environmental future.

Online resources designed to reduce climate anxiety, support resilience, development and empower individuals, such as online networks and climate cafés that gather participants from all over the world, are also worthy to mention in this context, even if empirical evidence about their emotional, motivational and behavioural effects are rather scarce.

Handlebar: Future Orientation in CCE

Recent research (e.g., Zhu, 2020) suggests that climate perception and mitigation are influenced by both national and personal differences in future orientation. Thus, countries with language that emphasizes the future tend to have higher levels of climate concern, and at the individual level future orientation helps people understand complex climate issues and make more personal mitigation decisions. Relying on such concepts, pro-environmental behavioural interventions and climate communications should be designed based on future orientation to be more effective. Specifically, intervention programs should include sets of skills necessary for envisioning desired futures, provide explanations on how different actors perceive and construct the future (Nalau and Cobb,

2022) and encourage multiple positive environmental visions (Pereira et al., 2020).

Prefigurative practices embedded into CCE are rooted in the simple and powerful idea that the future is always undecided and therefore open to change. Possible futuresmay be identified through young people's attitudes in coping with conflicts when asked about controversial climate issues (Ojala, 2022). Previous research has shown that future-focused individuals are more likely to care for the environment and take action to improve it (Milfont, Wilson and Diniz, 2012), to adopt conservation practices (Corral-Verdugo and Fraijo-Sing and Pinheiro, 2006), and to have pro-environmental beliefs (Milfont and Gouveia, 2006). The degree to which individuals are optimistic or pessimistic about the future of the environment is also an important factor in determining their environmental attitudes and actions.

Envisioning preferred and desirable futures is critical for increased environmental awareness and actions, especially through the integration of advanced technology, as suggested by Liu and Lin (2016) in a study with undergraduate science students. Based on the evidence, they advocate for the integration of future perspectives in CCE, as an effective strategy to encourage critical and creative thinking about the environmental and climate future, and, to also empower young people to adopt responsible behaviours for a sustainable future. Similarly, Lee et al. (2020) promote engagement in episodic future thinking regarding climate change-related risks as a trigger for higher risk perceptions and increased pro-environmental action, based on experimental results of their studies with undergraduates.

Climate change games, such as simulation-based and role-playing games, that combine interactive computer models with engaging scenarios, offer a promising approach to fostering future orientation in CCE (Pfirman et al., 2021; Meya and Eisenack, 2018). *Climate Change Simulation* and *World Climate Simulation* fall into the category of technology-enriched education and training frameworks for prefiguring the future and empowering actions for climate balance and sustainability, with deep cognitive, emotional and behavioural effects. Both games embed science-informed computer simulations and engaging role-playing. *Climate Change* is based on the interactive computer model *Energy Rapid Overview and Decision Support*, in short En-ROADS, developed by Climate Interactive, the MIT Sloan Sustainability Initiative, and Ventana Systems. It allows users to explore interactive energy and social transition scenarios, delivering immediate feedback on the impacts of policies and decisions related to energy supplies, greenhouse gas emissions, and expectations for climatic conditions in the 21st century (Rooney-Varga et al., 2020). Participants take on the roles of leaders from private companies, government and NGOs specializing in energy, climate policy/makers, etc. who can make decisions about their sector while trying to influence different group decisions as well as participating in a carbon pricing voting. Based on a mixed studyon pre and post-simulation surveys and focus groups, with 173 high school and university students organized into five groups of players, Rooney-Varga et al. (2020) concluded that the Climate Action Simulation improves participants' knowledge about emissions reductions and policies to mitigate climate change. Participants also reported deeper emotional engagement and a sense of empowerment in contributing to climate balance.

A similar interactive platform, *Climate Rapid Overview and Decision Support* or C-ROADS in brief, developed by the same research group together with UMPL Climate Change Initiative, is at the centre of World Climate Simulation, addressing the long-term impact of climate policy and actions in different world regions. The role-playing follows a similar scenario, with participants being challenged to negotiate as UN parties to create an international agreement that limits warming by 2100 to well below 2°C above preindustrial levels (Sterman et al., 2014). Rooney-Varga et al. (2018) analysed the outcomes of the World Climate Simulation resulting from more than 2,000 participants in 39 studies across eight countries with surveys conducted before, during and after the intervention. Three areas have shown statistically significant progress, namely knowledge of climate change causes, mechanisms and consequences, emotional engagement which includes a greater sense of urgency and hope, and the desire to learn more about climate change. The list of various climate change games constantly grows, but their design and impact at different levels and on different social groups need to be further studied, as researchers see that they may mainly attract informed and engaged players (Kwok, 2019) and oversimplify the climate change scenarios (Razali et al., 2022).

Discussion and Conclusion

Climate change has led to an increase in concerns about potential effective strategies and interventions that could be implemented to protect and conserve the environment. Even though various methods to combat climate change have been identified, there are still multiple difficulties in understanding and applying these methods in the educational context. CCE represents the cornerstone in the process of building a sustainable future. To increase the frequency of positive outcomes related to sustainable development, CCE could be implemented much more effectively with the help of new technologies, in a formal or informal manner (Monroe et al., 2019).

Starting from the Bicycle model on climate change education (Cantell et al., 2019), in this chapter we made a short synthesis of how we can improve and enrich CCE, with the help of ICT. Thus, this chapter offers a set of practical directions that can foster individual PEBs and that could be taken into account in the development of new educational curricula for climate change education programs or in developing new teaching strategies when addressing climate change. The chapter is structured into seven sections covering the component elements of a bicycle, which should work together to achieve the desired functionality: (1) wheels (i.e., knowledge and thinking skills); (2) frame (i.e., identity, values, and worldviews), (3) chains and pedals (i.e., actions to curb climate change); (4) saddle (i.e., motivation and participation); (5) brakes (i.e., operational barriers); (6) lamp (i.e., hope and other emotions); (7) handlebar (i.e., future orientation) (Cantell et al., 2019).

Regarding the transmission of knowledge related to climate change, there are multiple educational methods used in the literature (e.g., tutorials, applications that allow the calculation of the carbon footprint at an individual level, virtual trips that provide images of the substantial changes existing in certain regions or serious computer games). Developing knowledge through interactive, technology-based educational strategies helps students relate theoretical content learned to real, everyday life situations (Smith et al., 2006). The role of technologies in the assimilation of knowledge related to the environment is very well documented in the literature. However, to ensure a deep processing of information related to environmental issues and to develop connections between different disciplines or an exhaustive understanding of environmental problems, more strategies for developing critical thinking should be designed when using ICT systems. Along with a rich body of knowledge and skills related to critical thinking, the values and worldviews of individuals play a central role in achieving PEB. With the help of mobile applications that allow users to choose virtual identities that promote environmental protection and conservation, individual's biospheric values can be developed. In addition, the use of potential scenarios of environmental evolution can help change users' worldviews, but research on this matter is scarce. This issue is of particular interest to CCE, since worldview represents a significant obstacle in the adoption of climate change mitigation initiatives that would impact the lifestyle ofindividuals. Rooted in scientific skepticism, liberalism, anti-elitism, conspiracism, and political populism, such adverse worldviews would be hard to change (Lewandowsky, 2021). Social media (for debunking misinformation), realistic (versus apocalyptic) simulation, and virtual role-play can enhance information delivery, cultivation of responsibility and depolarization, and can cultivate a worldview that does not contrast with environmental policy in students, teachers, and the general population (O'Neill and Nicholson-Cole, 2009; Rooney-Varga et al., 2018; Van der Linden et al., 2017).

To ensure the informational transfer in a practical plan, other *ingredients* must be added. Thus, for good functionality of ICT strategies in educationa, technological tools must promote the formation of routines and *green habits*. In this way, we can ensure that the positive effects obtained through the use of mobile applications or gamification approaches will be long-term. For example, mobile applications should motivate users by highlighting real-world actions that they could perform to make a significant change. Also, keeping track of individual progress and highlighting others' progress, could foster social comparison and promote healthy social norms in virtual settings. Furthermore, for the individual to act and develop sustainable behavioural patterns, it is important to maintain an optimal level of intrinsic motivation (Van Der Linden, 2015). Mobile applications that use gamification or serious games can stimulate curiosity (e.g., collective problem-solving by generating innovative solutions, resolving virtual missions, and accepting different challenges) and can increase the intrinsic motivation of users to produce changes in the environment (Douglas and Brauer, 2021). Climate change games are useful and flexible tools to be used in the educational process. For example, games can increase the level of knowledge about environmental problems, foster the understanding of the complexity of climate systems, represent a binder for implementation in everyday life PEBs, and can motivate the individual because they contain various elements that reinforce desirable behaviours through rewards and virtual incentives (Ricoy and Sánchez-Martínez, 2022; Wu and Lee, 2015).

As for the operational barriers that might arise at the individual level (see Stern, 2000; Kollmuss and Agyeman, 2002; Gifford, 2011 for a review), it is necessary to identify them and bring them to students' attention through the CCE. Some barriers can be overcome using technological innovation and ICT (e.g., using IoT systems, nudging approaches, reducing perceived costs, or perceived temporal distance of climate change effects). To reduce perceived barriers, multi-level action strategies are needed that provide consistent visions to all actors involved in CCE (Quimby and Angelique, 2011). With the help of ICT, students can understand the utility of achieving PEB, teachers can effectively convey information and provide a complex framework related to environmental issues, and educational institutions could use ICT as an integrated strategy for sustainable institutional development. Finally, the emotions felt by those involved in the educational process can represent a driving force of CCE, and hope for a better future plays a central role (Ojala, 2023b). Collaborative-competitive games facilitate the understanding of future scenarios that are sustainable and that highlight the positive effects of green future alternatives. However, there is currently little data in the literature that indicates the role of ICT in the development of resilience to climate change and its relation to emotions such as eco-anxiety or eco-guilt. Using computer simulations that foreshadow a possible optimistic future and a positive trajectory of human action, can create a trajectory toward climate sustainability. By exploring some social transition scenarios, CCE fosters individual understanding of the impacts of policies and decision-making strategies, alongside a deeper emotional engagement. Overall, there is strong evidence that the use of technology in CCE can produce positive effects, both in the formal contexts (i.e., at school) and informal ones (i.e., at home), and the actions students can take in the virtual world could correspondingly affect the real world.

In conclusion, Information and Communication Technologies represent useful tools in delivering Climate Change Education. By implementing them in the holistic Bicycle model of CCE, we can respond to the need for participatory, interdisciplinary, creative, and affect-driven approaches that were highlighted by Rousell and Cutter-Mackenzie-Knowles (2020). Although currently more present in some components of the model than in others, the use of ICT can be continuously developed to respond to all the various needs of teachers and students. Moreover, such tools can be further implemented in the curriculum for different topics (not only in STEM education but also in social sciences and humanities) and in national and international educational policies and programs, while also representing viable resources for bottom-up, informal education initiatives.

References

Akhtar, S., Khan, K.U., Atlas, F. and Irfan, M. (2022). Stimulating student's pro-environmental behavior in higher education institutions: An ability–motivation–opportunity perspective. *Environment, Development and Sustainability*, *24*(3), 4128–4149. https://doi.org/10.1007/s10668-021-01609-4

Anderson, A. (2012). Climate change education for mitigation and adaptation. *Journal of Education for Sustainable Development*, *6*(2), 191–206. https://doi.org/10.1177/0973408212475199

Asimakopoulou, P., Nastos, P., Vassilakis, E., Hatzaki, M. and Antonarakou, A. (2021). Earth observation as a facilitator of climate change education in schools: The teachers' perspectives. *Remote Sensing*, *13*(8), 1587. https://doi.org/10.3390/rs13081587

Beasy, K., Jones, C., Kelly, R., Lucas, C., Mocatta, G., Pecl, G. and Yildiz, D. (2023). The burden of bad news: educators' experiences of navigating climate change education. *Environmental Education Research*, DOI: 10.1080/13504622.2023.2238136.

Bolderdijk, J.W., Gorsira, M., Keizer, K., Steg, L. (2013) Values Determine the (In) Effectiveness of Informational Interventions in Promoting Pro-Environmental Behavior. *PLoS ONE, 8*(12), e83911. https://doi.org/10.1371/journal.pone.0083911

Boncu, Ş., Candel, O.S. and Popa, N.L. (2022). Gameful green: a systematic review on the use of serious computer games and gamified mobile apps to foster pro-environmental information, attitudes and behaviors. *Sustainability*, *14*(16), 10400. https://doi.org/10.3390/su141610400

Boncu, Ş., Candel, O.S., Prundeanu, O. and Popa, N.L. (2023). Growing a digital iceberg for a polar bear: effects of a gamified mobile app on university students' pro-environmental behaviours. *International Journal of Sustainability in Higher Education*, *24*(8), 1932–1948. https://doi.org/10.1108/IJSHE-03-2023-0092

Brick, C., Bosshard, A. and Whitmarsh, L. (2021). Motivation and climate change: A review. *Current Opinion in Psychology*, *42*, 82–88. https://doi.org/10.1016/j.copsyc.2021.04.001

Cantell, H., Tolppanen, S., Aarnio-Linnanvuori, E. and Lehtonen, A. (2019). Bicycle model on climate change education: Presenting and evaluating a model. *Environmental Education Research*, *25*(5), 717–731. https://doi.org/10.1080/13504622.2019.1570487

Chusni, M.M., Saputro, S. and Rahardjo, S.B. (2021). Student's Critical Thinking Skills through Discovery Learning Model Using E-Learning on Environmental Change Subject Matter. *European Journal of Educational Research*, *10*(3), 1123–1135. https://doi.org/10.12973/eu-jer.10.3.1123

Cook, J.W. (2019). Learning at the edge of history. pp. 1–28 *In*: J.W. Cook (ed.). *Sustainability, human well-being, and the future of education*. Palgrave Macmillan.

Corral-Verdugo, V., Fraijo-Sing, B. and Pinheiro, J.Q. (2006). Sustainable Behavior and Time Perspective: Present, Past, and Future Orientations and their Relationship with Water Conservation Behavior. *Interamerican Journal of Psychology*, 40(2), 139–147.

Cox, H., Kelly, K. and Yetter, L. (2014). Using remote sensing and geospatial technology for climate change education. *Journal of Geoscience Education*, *62*(4), 609–620. https://doi.org/10.5408/13-040.1

Douglas, B.D. and Brauer, M. (2021). Gamification to prevent climate change: A review of games and apps for sustainability. *Current Opinion in Psychology, 42*, 89–94. https://doi.org/10.1016/j.copsyc.2021.04.008

Dreijerink, L., Handgraaf, M. and Antonides, G. (2022). The impact of personal motivation on perceived effort and performance of pro-environmental behaviors. *Frontiers in Psychology, 13*, 977471. https://doi.org/10.3389/fpsyg.2022.977471

Fernández Galeote, D., Legaki, N.Z. and Hamari, J. (2022). Avatar identities and climate change action in video games: analysis of mitigation and adaptation practices. In: *Proceedings of the 2022 CHI Conference on Human Factors in Computing Systems* (pp. 1-18). https://doi.org/10.1145/3491102.3517438

Fernández Galeote, D., Rajanen, M., Rajanen, D., Legaki, N.Z., Langley, D.J. and Hamari, J. (2021). Gamification for climate change engagement: review of corpus and future agenda. *Environmental Research Letters, 16*(6), 063004. DOI 10.1088/1748-9326/abec05

Finnegan, W. (2023). Educating for hope and action competence: a study of secondary school students and teachers in England. *Environmental Education Research, 29*(11), 1617–1636. https://doi.org/10.1080/13504622.2022.2120963

Geiger, N., Dwyer, T. and Swim, J.K. (2023). Hopium or empowering hope? A meta-analysis of hope and climate engagement. *Frontiers in Psychology*, 14, 1139427. DOI:10.3389/fpsyg.2023.1139427.

Geiger, N., Gasper, K., Swim, J.K. and Fraser, J. (2019). Untangling the components of hope: increasing pathways (not agency) explains the success of an intervention that increases educators' climate change discussions. Journal of Environmental Psychology, 66, 101366. DOI: 10.1016/j.jenvp.2019.101366.

Gifford, R. (2011). The dragons of inaction: Psychological barriers that limit climate change mitigation and adaptation. *American Psychologist, 66*(4), 290–302. https://doi.org/10.1037/a0023566

Gold, A.U., Kirk, K., Morrison, D., Lynds, S., Sullivan, S.B., Grachev, A. and Persson, O. (2015). Arctic climate connections curriculum: A model for bringing authentic data into the classroom. *Journal of Geoscience Education, 63*(3), 185-197. https://doi.org/10.5408/14-030.1

Gschwandtner, A., Jewell, S. and Kambhampati, U.S. (2022). Lifestyle and life satisfaction: the role of delayed gratification. *Journal of Happiness Studies, 23*(3), 1043–1072. https://doi.org/10.1007/s10902-021-00440-y

Hornsey, M.J. (2021). The role of worldviews in shaping how people appraise climate change. *Current Opinion in Behavioral Sciences, 42*, 36–41. https://doi.org/10.1016/j.cobeha.2021.02.021

Hickman, C., Marks, E., Pihkala, P., Clayton, S., Lewandowski, E., Mayall, E., Wray, B., Mellor, C. and van Susteren, L. (2021). Climate anxiety in children and young people and their beliefs about government responses to climate change: a global survey. *Lancet Planetary Health*, 5(12), e863–e873. DOI:10.1016/S2542-5196(21)00278-3.

Huang, L., Wen, Y. and Gao, J. (2020). What ultimately prevents the pro-environmental behavior? An in-depth and extensive study of the behavioral costs. *Resources, Conservation and Recycling, 158*, 104747. https://doi.org/10.1016/j.resconrec.2020.104747

Inzlicht, M., Shenhav, A. and Olivola, C.Y. (2018). The effort paradox: Effort is both costly and valued. *Trends in Cognitive Sciences, 22*(4), 337–349. https://doi.org/10.1016/j.tics.2018.01.007

Janakiraman, S., Watson, S.L., Watson, W.R. and Newby, T. (2021). Effectiveness of digital games in producing environmentally friendly attitudes and behaviors: A mixed methods study. *Computers & Education*, 160, 104043, https://doi.org/10.1016/j.compedu.2020.104043.

Judah, G., Gardner, B. and Aunger, R. (2013). Forming a flossing habit: An exploratory study of the psychological determinants of habit formation. *British Journal of Health Psychology*, *18*(2), 338–353. https://doi.org/10.1111/j.2044-8287.2012.02086.x

Karpudewan, M. and Mohd Ali Khan, N.S. (2017). Experiential-based climate change education: Fostering students' knowledge and motivation towards the environment. *International Research in Geographical and Environmental Education*, *26*(3), 207–222. https://doi.org/10.1080/10382046.2017.1330037

Kantenbacher, J., Miniard, D., Geiger, N., Yoder, L. and Attari, S.Z. (2022). Young adults face the future of the United States: perceptions of its promise, perils, and possibilities. *Futures*, *139*, 102951. DOI: 10.1016/j.futures.2022.102951.

Kerret, D., Orkibi, H. and Ronen, T. (2016). Testing a model linking environmental hope and self-control with students' positive emotions and environmental behaviour. *The Journal of Environmental Education*. 47(4), 307–317. DOI: 10.1080/00958964.2016.1182886.

Keuschnigg, M. and Kratz, F. (2018). Thou shalt recycle: How social norms of environmental protection narrow the scope of the low-cost hypothesis. *Environment and Behavior*, *50*(10), 1059–1091. https://doi.org/10.1177/0013916517726569

Kleinlogel, E.P., Mast, M.S., Renier, L.A., Bachmann, M. and Brosch, T. (2023). Immersive virtual reality helps to promote pro-environmental norms, attitudes and behavioural strategies. *Cleaner and Responsible Consumption*, *8*, 100105. https://doi.org/10.1016/j.clrc.2023.100105

Knol, E. and De Vries, P.W. (2011). EnerCities, a serious game to stimulate sustainability and energy conservation: Preliminary results. *eLearning Papers*, 25, 1–10. https://papers.ssrn.com/sol3/papers.cfm?abstract_id=1866206

Kollmuss, A. and Agyeman, J. (2002). Mind the gap: why do people act environmentally and what are the barriers to pro-environmental behavior?. *Environmental education research*, *8*(3), 239–260. https://doi.org/10.1080/13504620220145401

Kolokotsa, D., Gobakis, K., Papantoniou, S., Georgatou, C., Kampelis, N., Kalaitzakis, K. and Santamouris, M. (2016). Development of a web based energy management system for University Campuses: The CAMP-IT platform. *Energy and Buildings*, *123*, 119-135. https://doi.org/10.1016/j.enbuild.2016.04.038

Kumar, P., Sahani, J., Rawat, N., Debele, S., Tiwari, A., Emygdio, A.P.M. and Pfautsch, S. (2023). Using empirical science education in schools to improve climate change literacy. *Renewable and Sustainable Energy Reviews*, *178*, 113232. https://doi.org/10.1016/j.rser.2023.113232

Kwok, R. (2019). Can climate change games boost public understanding?. *Proceedings of the National Academy of Sciences*, *116*(16), 7602–7604. DOI:10.1073/pnas.1903508116

Lawrance, E.L., Thompson, R., Newberry Le Vay, J., Page, L. and Jennings, N. (2022). The impact of climate change on mental health and emotional wellbeing: a narrative review of current evidence, and its implications. *International Review of Psychiatry*, *34*(5), 443–498. https://doi.org/10.1080/09540261.2022.2128725

Leal Filho, W., Sima, M., Sharifi, A., Luetz, J.M., Salvia, A.L., Mifsud, M. and Lokupitiya, E. (2021). Handling climate change education at universities: an overview. *Environmental Sciences Europe*, *33*, 1–19. https://doi.org/10.1186/s12302-021-00552-5

Lee, J.J., Ceyhan, P., Jordan-Cooley, W. and Sung, W. (2013). GREENIFY: A real-world action game for climate change education. *Simulation & Gaming*, *44*(2–3), 349–365. https://doi.org/10.1177/1046878112470539

Lee, P.S., Sung, Y.H., Wu, C.C., Ho, L.C. and Chiou, W.B. (2020). Using episodic future thinking to pre-experience climate change increases pro-environmental behavior. *Environment and Behavior*, *52*(1), 60–81. DOI: 10.1177/0013916518790590.

Lewandowsky, S. (2021). Climate change disinformation and how to combat it. *Annual Review of Public Health*, *42*, 1–21. https://doi.org/10.1146/annurev-publhealth-090419-102409

Liu, S.C. and Lin, H.S. (2016). Envisioning preferred environmental futures: exploring relationships between future-related views and environmental attitudes. *Environmental Education Research*, DOI:10.1080/13504622.2016.1180504.

Liu, P., Teng, M. and Han, C. (2020). How does environmental knowledge translate into pro-environmental behaviors?: The mediating role of environmental attitudes and behavioral intentions. *Science of the Total Environment*, *728*, 138126. https://doi.org/10.1016/j.scitotenv.2020.138126

Mejía-Cáceres, M.A., Rieckmann, M. and Folena Araújo, M.L. (2023). Political Discourses as A Resource for Climate Change Education: Promoting Critical Thinking by Closing the Gap between Science Education and Political Education. *Sustainability*, *15*(8), 6672. https://doi.org/10.3390/su15086672

Meya, J. and Eisenack, K. (2018). Effectiveness of gaming for communicating and teaching climate change. *Climactic Change*. 149 (3–4), 319–333. DOI:10.1007/s10584-018-2254-7.

Milfont, T.L. and V.V. Gouveia (2006). Time perspective and values: an exploratory study of their relations to environmental attitudes. *Journal of Environmental Psychology*, *26*(1), 72–82.

Milfont, T.L., J. Wilson and P. Diniz (2012). Time perspective and environmental engagement: a meta-analysis. *International Journal of Psychology*, 47(5), 325–334. http://dx.doi.org/10.1080/00207594.2011.647029.

Monroe, M.C., Plate, R.R., Oxarart, A., Bowers, A. and Chaves, W.A. (2019). Identifying effective climate change education strategies: A systematic review of the research. *Environmental Education Research*, *25*(6), 791–812. https://doi.org/10.1080/13504622.2017.1360842

Mochizuki, Y. and Bryan, A. (2015). Climate change education in the context of education for sustainable development: Rationale and principles. *Journal of Education for Sustainable Development*, *9*(1), 4–26. https://doi.org/10.1177/0973408215569109

Nalau, J. and Cobb, G. (2022). The strengths and weaknesses of future visioning approaches for climate change adaptation: A review. *Global Environmental Change*, 74, 102527. https://doi.org/10.1016/j.gloenvcha.2022.102527.

Ochmann, J. and Lehrer, C. (2023). One Save per Day: How Mobile Technology can Support Individuals to Adopt Pro-environmental Behaviors. pp. 761–770. *In*: *The 56th Hawaii International Conference on System Sciences. HICSS 2023*. Hawaii International Conference on System Sciences (HICSS). https://hdl.handle.net/10125/102725

Ojala, M. (2022). Prefiguring sustainable futures? Young people's strategies to deal with conflicts about climate-friendly food choices and implications for transformative learning. *Environmental Education Research*, 28(8), 1157–1174, DOI:10.1080/13504622.2022.2036326.

Ojala, M. (2023a). Climate-change education and critical emotional awareness (CEA): Implications for teacher education. *Educational Philosophy and Theory*, 55(10), 1109-1120, DOI: 10.1080/00131857.2022.2081150.

Ojala, M. (2023b). Hope and climate-change engagement from a psychological perspective. *Current Opinion in Psychology, 49*, 101514. https://doi.org/10.1016/j.copsyc.2022.101514

O'Neill, S. and Nicholson-Cole, S. (2009). "Fear won't do it" promoting positive engagement with climate change through visual and iconic representations. *Science communication, 30*(3), 355–379. https://doi.org/10.1177/10755470083292

Pereira, L.M., Davies, K.K., Belder, E., Ferrier, S., Karlsson-Vinkhuyzen, S., Kim, H., Kuiper, J. J., Okayasu, S., Palomo, M.G., Pereira, H.M., Peterson, G., Sathyapalan, J., Schoolenberg, M., Alkemade, R., Carvalho Ribeiro, S., Greenaway, A., Hauck, J., King, N., Lazarova, T. and Lundquist, C.J. (2020). Developing multiscale and integrative nature-people scenarios using the nature futures framework. *People and Nature*, 2, 1172-1195. https://doi.org/10.1002/pan3.10146.

Petersen, G.B., Klingenberg, S., Mayer, R.E. and Makransky, G. (2020). The virtual field trip: Investigating how to optimize immersive virtual learning in climate change education. *British Journal of Educational Technology, 51*(6), 2099–2115. https://doi.org/10.1111/bjet.12991

Peterson, C. and Seligman, M.E. (2004). *Character Strengths and Virtues: A Handbook and Classification*, Vol. 1. New York, NY: Oxford University Press.

Pfirman, S., O'Garra, T., Bachrach, S.E., Brunacini, J., Reckien, D., Lee, J.J and Lukasiewicz, E. (2021). "Stickier" learning through gameplay: An effective approach to climate change education. *Journal of Geoscience Education. 69*(2), 192–206. DOI:10.1080/10899995.2020.1858266.

Plechatá, A., Morton, T., Perez-Cueto, F.J. and Makransky, G. (2022). Why just experience the future when you can change it: Virtual reality can increase pro-environmental food choices through self-efficacy. *Technology, Mind, and Behavior, 3*(4), 1–12. https://doi.org/10.1037/tmb0000080

Pruneau, D., Doyon, A., Langis, J., Vasseur, L., Ouellet, E., McLaughlin, E. and Martin, G. (2006). When teachers adopt environmental behaviors in the aim of protecting the climate. *The Journal of Environmental Education, 37*(3), 3–12. https://doi.org/10.3200/JOEE.37.3.3-12

Quimby, C.C. and Angelique, H. (2011). Identifying barriers and catalysts to fostering pro-environmental behavior: Opportunities and challenges for community psychology. *American Journal of Community Psychology, 47*, 388–396. https://doi.org/10.1007/s10464-010-9389-7

Radziszewski, K., Anacka, H., Obracht-Prondzyńska, H., Tomczak, D., Wereszko, K. and Weichbroth, P. (2021). Greencoin: prototype of a mobile application facilitating and evidencing pro-environmental behavior of citizens. *Procedia Computer Science, 192*, 2668–2677. https://doi.org/10.1016/j.procs.2021.09.037

Razali, N.E.M., Ramli, R.Z., Mohamed, H., Zin, N.a.M., Rosdi, F. and Diah, N.M. (2022). Identifying and validating game design elements in serious game guideline for climate change. *Heliyon*, 8(1), e08773. DOI:10.1016/j.heliyon.2022.e08773.

Ricoy, M.C. and Sánchez-Martínez, C. (2022). Raising ecological awareness and digital literacy in primary school children through gamification. *International Journal of*

Environmental Research and Public Health, 19(3), 1149. https://doi.org/10.3390/ijerph19031149

Ro, M., Brauer, M., Kuntz, K., Shukla, R. and Bensch, I. (2017). Making Cool Choices for sustainability: Testing the effectiveness of a game-based approach to promoting pro-environmental behaviors. *Journal of Environmental Psychology, 53*, 20–30. https://doi.org/10.1016/j.jenvp.2017.06.007

Rooney-Varga, J.N., Kapmeier, F., Sterman, J.D., Jones, A.P., Putko, M. and Rath, K. (2020). The Climate Action Simulation. *Simulation & Gaming*, 51(2). 114–140. https://doi.org/10.1177/1046878119890643.

Rooney-Varga, J.N., Sterman, J.D., Fracassi, E., Franck, T., Kapmeier, F., Kurker, V., Johnston, E., Jones, A.P. and Rath, K. (2018). Combining role-play with interactive simulation to motivate informed climate action: Evidence from the World Climate simulation. *PLoS ONE*, 13(8), e0202877. https://doi.org/10.1371/journal.pone.0202877.

Ross, H., Rudd, J.A., Skains, R.L. and Horry, R. (2021). How big is my carbon footprint? Understanding young people's engagement with climate change education. *Sustainability, 13*(4), 1961. https://doi.org/10.3390/su13041961

Rousell, D. and Cutter-Mackenzie-Knowles, A. (2020). A systematic review of climate change education: Giving children and young people a 'voice'and a 'hand'in redressing climate change. *Children's Geographies, 18*(2), 191–208. https://doi.org/10.1080/14733285.2019.1614532

Rousell, D., Wijesinghe, T., Cutter-Mackenzie-Knowles, A. and Osborn, M. (2023). Digital media, political affect, and a youth to come: Rethinking climate change education through Deleuzian dramatisation. *Educational Review, 75*(1), 33–53. https://doi.org/10.1080/00131911.2021.1965959

Sarewitz, D. (2011). Does climate change knowledge really matter?. *Wiley Interdisciplinary Reviews: Climate Change, 2*(4), 475–481. https://doi.org/10.1002/wcc.126

Soliman, M., Alisat, S., Bashir, N.Y. and Wilson, A.E. (2018). Wrinkles in time and drops in the bucket: Circumventing temporal and social barriers to pro-environmental behavior. *Sage Open, 8*(2), 2158244018774826. https://doi.org/10.1177/2158244018774826

Smith, J.M., Edwards, P.M. and Raschke, J. (2006). Using technology and inquiry to improve student understanding of watershed concepts. *Journal of Geography, 105*(6), 249–257. https://doi.org/10.1080/00221340608978694

Steele, J. (2020). What is (perception of) effort? Objective and subjective effort during task performance. *PsyArXiv*. [Preprint]. doi: 10.31234/osf.io/kbyhm

Stenberdt, V.A. and Makransky, G. (2023). Mastery experiences in immersive virtual reality promote pro-environmental waste-sorting behavior. *Computers & Education, 198*, 104760. https://doi.org/10.1016/j.compedu.2023.104760

Sterman, J.D., Franck, T., Fiddaman, T., Jones, A., McCauley, S., Rice, P., Sawin, E., Siegel, L. and Rooney-Varga, J.N. (2014). WORLD CLIMATE: A Role-Play Simulation of Climate Negotiations. Simulation & Gaming, 46(3–4), 348–382. https://doi.org/10.1177/1046878113514935.

Stern, P.C. (2000). New environmental theories: toward a coherent theory of environmentally significant behavior. *Journal of Social Issues, 56*(3), 407–424. https://doi.org/10.1111/0022-4537.00175

Stevenson, K.T., Peterson, M.N., Bondell, H.D., Moore, S.E. and Carrier, S.J. (2014). Overcoming skepticism with education: interacting influences of worldview and climate change knowledge on perceived climate change risk among adolescents. *Climatic change*, *126*, 293–304. https://doi.org/10.1007/s10584-014-1228-7

Suwatra, W., Suyatna, A. and Rosidin, U. (2018). Development of interactive e-module for global warming to grow of critical thinking skills. *International Journal of Advanced Engineering, Management and Science*, *4*(7), 264307. https://dx.doi.org/10.22161/ijaems.4.7.7

Tabuenca, B., García-Alcántara, V., Gilarranz-Casado, C. and Barrado-Aguirre, S. (2020). Fostering environmental awareness with smart IoT planters in campuses. *Sensors*, *20*(8), 2227. https://doi.org/10.3390/s20082227

Taghikhah, F., Raffe, W.L., Mitri, G., Du Toit, S., Voinov, A. and Garcia, J.A. (2019). Last Island: Exploring Transitions to Sustainable Futures through Play. *ACSW '19: Proceedings of the Australasian Computer Science Week Multiconference January 2019*. Article No.: 41. pp. 1–7. https://doi.org/10.1145/3290688.3290746.

Tolppanen, S., Cantell, H., Aarnio-Linnanvuori, E. and Lehtonen, A. (2017). Pirullisen ongelman äärellä—kokonaisvaltaisen ilmastokasvatuksen malli [Dealing with a wicked problem—a model for holistic climate change education]. *Kasvatus, 5*, 456–468.

Tolppanen, S., Kang, J. and Riuttanen, L. (2022). Changes in students' knowledge, values, worldview, and willingness to take mitigative climate action after attending a course on holistic climate change education. *Journal of Cleaner Production*, *373*, 133865. https://doi.org/10.1016/j.jclepro.2022.133865

Van Der Linden, S. (2015). Intrinsic motivation and pro-environmental behavior. *Nature Climate Change*, *5*(7), 612–613. https://doi.org/10.1038/nclimate2669

Van der Linden, S., Maibach, E., Cook, J., Leiserowitz, A. and Lewandowsky, S. (2017). Inoculating against misinformation. *Science*, *358*(6367), 1141–1142. DOI: 10.1126/science.aar4533

Wu, J.S. and Lee, J.J. (2015). Climate change games as tools for education and engagement. *Nature Climate Change*, *5*(5), 413–418. https://doi.org/10.1038/nclimate2566

Yee, N., Bailenson, J.N. and Ducheneaut, N. (2009). The Proteus effect: Implications of transformed digital self-representation on online and offline behavior. *Communication Research*, *36*(2), 285–312. https://doi.org/10.1177/00936502083302

Zhu, J., Hu, S., Wang, J. and Zheng, X. (2020). Future orientation promotes climate concern and mitigation. *Journal of Cleaner Production*, 262, 121212. https://doi.org/10.1016/j.jclepro.2020.121212.

CHAPTER 7

Integration of Climate Change and Green Technologies into Language Education

*Yunus Emre Akbana** and *Stefan Rathert*

Introduction

There is no doubt that climate change is among the most significant issues of our times, having considerable impact not only locally on individuals' lives but also on the future of the Earth (Dahl, 2011). Consequently, there is urgent need for action to be taken by authorities and the public to mitigate the effects of climate change (Arıkan and Günay, 2021). One of the primary tasks is to set educational policies conducive to raising awareness of climate change and the potential of green technologies, beginning at an early age. Moreover, there is the need to initiate professional development for teachers, so that they can integrate climate change into their instructional practices (Boon, 2010; Chopra et al., 2019).

The immense impact and the complexity of climate change necessitates the development of cross-discipline curricula to deal with the topic, encompassing different subject areas ranging from the natural

Kahramanmaraş Sütçü İmam University, Türkiye.
Email: srathert@ksu.edu.tr
* Corresponding author: yakbana@ksu.edu.tr

sciences to social sciences and humanities (Anderson, 2012). Indeed, there are educational programmes attempting to disseminate content and practices to raise learners' awareness of climate change. The UNESCO *Climate Change Education for Sustainable Development* programme, for example, seeks to help citizens "understand, address, mitigate and adapt to the impacts of climate change, encourage the changes in attitudes and behaviours needed to (...) build a new generation of climate change-aware citizens" (UNESCO, 2010, p. 4). *Sandwatch*, another UNESCO programme launched in 1998, is a multidisciplinary programme bridging science with the humanities and offering extracurricular activities to enrich classroom-based instruction (Chopra et al., 2019). The *Trans-Disciplinary Research Oriented Pedagogy for Improving Climate Studies and Understanding*, a global project funded by the International Science Council, provides educational resources to integrate climate change into different school subjects through "interactive visualisations, models and simulators, classroom/laboratory activities, online readings, games, mobile applications, and video micro-lectures" (Chopra et al., 2019, p. 58). Although this project does not target language education, it provides comprehensive guidelines on how to engage in multidisciplinary education. Through its multimodal texts (written texts, images, audios, videos) integrated in a cognitively and affectively challenging sequence of activities, the project exposes learners to the target language and invites them to negotiate the significance of climate change and technology with their peers. This way the project facilitates language learning and raises learner awareness of climate change.

These three programmes indicate that the topic of climate change has found its way into educational practices. Foreign language teaching and learning, even though not first and foremost associated with science, offers opportunities to integrate content surrounding climate change and technology for two major reasons: First, language must be taught over some sort of content, and the climate change and technology as topics present in media and public discourse are potentially engaging for learners and teachers. Second, educational technologies are potentially environmentally friendly (e.g., they minimise the use of photocopies or travelling to locations of instruction). However, the integration of climate change and technology in foreign language teaching is currently an under-researched area. Programme developers design curricula and materials that foresee the integration of climate change and technology (Akbana and Yavuz, 2022; Stibbe, 2004), but the extent to which they are considered in classrooms depends on how teachers and learners appropriate them.

To explore this aspect, this case study reports on the views of English language teachers, preservice teachers and learners on the integration of climate change and green technology in the teaching of English as a foreign language (EFL). An additional asset of this study is its research context: Türkiye as a middle-income country has a population whose awareness of the impact of climate change and the role of technology is arguably lower compared to high-income countries (Arıkan and Günay, 2021). Examining this particular may offer insights in how to promote foreign language teaching that is informed by eco-pedagogy, which aims at enabling learners to reflect on environmental issues and act accordingly, and ecolinguistics, which aims at enabling learners to understand and engage in discourse about environmental issues (e Zia et al., 2023).

The following two sections present a survey on the integration of climate change and technology in foreign language materials and the views teachers hold on this content. Materials, particularly coursebooks, significantly shape instructional practice as they present teachers and learners the content over which the language is taught and learned, often wavering between contemporary issues and avoiding controversial topics (Gray, 2016). Indeed, material design and views on educational practice are overlapping areas as experiences with materials shape the views of those who use them (McGrath, 2013).

Climate Change and Technology in English Language Coursebooks

In line with the significance of climate change and technology in the public discourse, related content has been well considered by material designers. Studies conducted in different contexts, however, draw a mixed picture on the extent to which such content is considered. Yılmaz and Güleryüz Adamhasan (2022), for example, examining the inclusion of global issues in global EFL coursebooks, showed that several global topics ranging from environmental issues to health and socioeconomic conditions are present in the materials. With specific reference to climate change, the authors report on a text that discusses governmental investments in technology to respond to climate change, which, as the authors conclude, demonstrates that coursebooks potentially guide students towards a sustainable lifestyle. To give another example on technology, the coursebooks contain texts that inform students on the use of alternative energy sources and recycling, thereby displaying the writers' awareness of the potential technology affords to address ecological challenges.

Jodoin and Singer (2018) report that learners are only moderately exposed to environmental topics in materials used in Japanese higher education. Based on a corpus analysis of more than 55.000 words in EFL coursebooks, the study suggests that materials only slightly promote sustainable development because the presentation of topics like climate change does not translate to meaningful critical examination where learners question their beliefs, develop new understandings and eventually change their behaviour.

Rather sporadic consideration of eco-pedagogical and eco-linguistic approaches was also found in local coursebooks used in Iranian schools with topics restricted to themes such as weather, saving nature, and renewable energy (Faramarzi and Janfeshan, 2021). In a similar vein, Mohammadnia and Moghadam (2019) found in their examination of an EFL coursebook series for adults that climate change as a topic is mentioned only three times among other global issues. Two studies from Indonesia differ from the findings revealed in the Iranian context as 20 percent of the content of an EFL coursebook series for adolescents is related to climate change, waste management, nature conservation, wildlife conservation, sustainable fishing, and environmental and social balance (Inayati et al., 2016). The topic of climate change is also present in the materials scrutinised by Teridinanti et al. (2021), but the topic is displayed in expository texts that are unlikely to offer learners the opportunity to critically approach the thematic content and challenge their own behaviour.

Teacher, Preservice Teacher and Learner Views on the Integration of Climate Change and Technology

The existence or the lack of climate change and technical solutions in language learning materials does not guarantee their integration into instructional practice. Teachers may skip such content or focus on linguistic aspects so that the topic is not discussed but functions as a vehicle to teach grammar or vocabulary, or they may add their own materials because they have particular sensitivities or aversions towards environmental and technological issues. In any case, it is important to explore teacher and learner beliefs and attitudes towards climate change to design professional development programmes and instructional practices conducive to integrating climate change and green technology into language teaching (Boon, 2014).

Studies report on diverse teacher and learner views. Yasukawa (2023), for instance, investigated English language and literacy teachers' views of integrating climate change and other environmental issues into their teaching in a vocational college in Australia. Some teachers favoured integrating such topics into literacy teaching, while others–neglecting curriculum requirements–avoided integrating environmental issues as they believed they were better able to meet learner needs and interests this way. Other teachers added extra materials (e.g., TED talks) to raise their learners' awareness of the dangers of climate change and the potential of technological advancement and inspired other teachers to integrate similar content into their teaching.

Teachers who are willing to integrate global issues like climate change into their teaching may be hindered by unsuitable teaching materials or curricular requirements. e Zia et al. (2023), for example, report on Pakistani teachers who feel desperate because their materials insufficiently cover related topics and the state authorities do not provide professional development to train teachers to deal with climate change and technology in their EFL classes. Apart from a lack of pedagogical knowledge, i.e., how to transform environment-related content into classroom activities, unfamiliarity with climate change and green technology per se has been reported as a hindrance for preservice and in-service teachers to integrate content into their lessons (e.g., Ramadhan et al., 2019). Lack of knowledge accompanied by a mistrust of climate change sources may additionally exert detrimental effects (Boon, 2014).

An examination of Turkish preservice EFL teachers' attitudes and tendencies towards including environmental education into classroom teaching revealed moderately positive attitudes and intentions to follow such an integration (Gürsoy and Sağlam, 2011). Akbana and Yavuz (2020) found that EFL teachers working at university English preparatory programmes had developed an awareness conducive to engaging in environmental peace education by integrating climate change in their language teaching. Some teachers, however, claimed that the extent the coursebooks offered related content restricted them, while other teachers included global warming through extracurricular activities. Another study conducted in Turkey found that fourth-year preservice EFL teachers and secondary school students favoured incorporating environmental peace education into teaching grammar as part of the English language instruction (Arıkan, 2009).

Gürsoy (2010) argues that if environmental education becomes a part of foreign language lessons in primary schools, young learners can both gain language proficiency and develop environmental awareness. Apparently, global issues attract learner interest in different age groups (Bayraktar Balkır, 2021), and the learning experience is impoverished when learners suffer from a lack of suitable materials (Amjad et al., 2022). This is unfortunate given the potential of integrating climate change and green technology to contribute to a transformative pedagogy that enables learners to develop "an eco-ethical consciousness consonant with their cultural religious identity expression" (Goulah, 2017, p. 109).

Methodology

The study

The integration of climate change and green technology into the teaching and learning of foreign languages is justified to raise learner awareness of environmental issues and to enrich instructional practice by covering relevant and potentially engaging content. However, how the integration is viewed by foreign language teachers, preservice teachers and learners is not well researched. Therefore, the study sought to answer the following research questions:

1. What do climate change and technology mean to the study participants?
2. What are their views and experiences of integrating climate change and technology into language teaching and learning?
3. Do the participants consider educational technology to be conducive to addressing climate change in language education?

Study Context and Participants

Attempting to examine the views of participants in a local context, English language instructors, EFL preservice teachers and learners (who would continue as EFL preservice teachers after passing the programme) at a Turkish state university were invited to participate in the study. This means all participants were associated with English language teaching as it was or would be their profession. The preservice teachers were at the intersection of language learning and teacher education, and it was assumed that the learners would draw on their ongoing experience of language learning. Eleven instructors (mean age = 43,7), 33 preservice teachers (mean age = 21,7), and 26 learners (mean age = 19,1) responded to the survey.

Data Collection

After obtaining ethical approval, data was collected through open-ended survey questions formulated in Turkish. To develop the semi-structured survey guide, the researchers generated questions considering the different target groups; the surveys consisted of both questions asked to all participants and questions tailored to the specific groups. All participants were asked about their understanding of climate change, how technology could address climate change and if this content should be considered for use in language classrooms. The instructors were asked about their own instructional practices of integrating climate change and technology or their reasons for avoiding such kind of teaching. The preservice teachers and learners were asked about their own learning experiences. The preservice teachers additionally were asked if they would integrate climate change and technology into their future teaching. A final question directed to all participants aimed at getting their views on how educational technology could contribute to slowing down or stopping climate change.

Piloting procedures included the presentation of questions to colleagues in the faculty of education to receive expert opinions, and to members of the target groups who would not take part in the survey. After incorporating modifications, the questions were given to an expert in Turkish language to finalise the piloting.

Based on preliminary results coming from the survey data, questions for oral interviews were developed following similar piloting procedures as for the survey. Participants who stated their willingness to be interviewed orally attended the interviews. Applying random sampling, three volunteers from each group took part in the oral interviews. The interview sessions were run in Turkish and each lasted around 20 minutes in Zoom and face-to-face sessions. The interviews were recorded and transcribed verbatim.

Data Analysis

The survey and interview data were analysed with content analysis. An inductive approach was followed. Both researchers engaged in repeated reading through the answers to get familiar with the data and then codified them employing structural coding, i.e., with a focus on the research questions (Saldana, 2009). After the first individual coding, the researchers compared their codes and developed a coding manual with operational definitions. This procedure, individual coding and comparing the coding results was repeated for the rest of the data. This way, inter-rater reliability was cared for and codes were tested resulting in re-coding when new codes emerged or codes were merged.

Results

The results are presented here in subsections that document the themes emerging as an outcome of the data analysis. Excerpts from the data illustrate the perceptions the participants hold. References in brackets refer to instructor (I), preservice teachers (PR) and learners (L), and participant ID numbers.

Conceptualizations of Climate Change

The participants in all groups mostly associated climate change with its adverse effects. Such effects were specifically named by some participants, such as "seasonal shifts, the danger of droughts and water shortage, and forced migration" (I9). The danger of extinction was more frequently expressed by the preservice teachers (33%) and learners (27%), while only one instructor pointed to the life-threatening danger of climate change. Only a few participants (three instructors, one preservice teacher and two learners) stated that climate change was human-caused, while one participant in each group said that climate change did not have a meaning for them or had no effect. In the interviews, a learner stated that "climate change is serious but not for me because it will affect the world three generations later" (L1), while an instructor blamed mankind: "We disrupt the balance of nature and nature has already started to take the revenge from us by climate change" (I2).

Perceived Roles of Technology to Address Climate Change

The responses showed that the participants did not have specialist knowledge of technology and how to associate it with climate change. Two preservice teachers and five learners explicitly stated that they had no idea how technology could be associated with climate change: "As technology is not an area I'm interested in, I don't believe that I have sufficient expertise. However, when I try to consider today's conditions, certain measurement devices and other equipment may be of use" (PR19). This voice points to the role of technology in examining climate change. However, this function was identified as a potential role by only one instructor and three preservice teachers. Much more prominent was the understanding of technology as a means to address problems caused by climate change (mentioned by 64% of the instructors, 39% of the preservice teachers and 23% of the learners) and to provide solutions to climate change (mentioned by 64%, 48% and 35%, respectively). The participants showed rather simplistic understandings of technology as they either pointed to its potential to provide solutions for global warming or its negative impacts as displayed in this excerpt: "Less technology is crucial to diminish climate change because advancement in technology is one of the main reasons for the increase in carbon dioxide emission" (L9). On the other hand, one instructor gave an example on the facilitative role of technology: "A local textile company uses technology to recycle and reuse the water in the factory, now they save 90 percent water and use only 10 percent for their needs in the production line" (I2). Still, a discussion of the ambivalent role of technology in climate change was not frequent in the data, and technology was scarcely recognized as conducive to researching climate change.

Views on Integrating Climate Change and Technology into Language Teaching

Asked if climate change and technology should be part of foreign language teaching, the majority of participants answered in the affirmative: Ninety-one percent of the instructors, 88 percent of the preservice teachers and 73 percent of the learners said that the content should be integrated. The data displayed participants' reflections on the appropriateness of climate change and technology as in this excerpt:

> This is a very important issue to reflect on. It may look attractive to integrate it [climate change and technology] into language teaching, but I don't think that the students will care about the content because the focus is not on dealing with such topics but on learning a language. (PR2)

While this teacher candidate, with an appreciation of the content's significance, denied the place of climate change and technology in language teaching, some preservice teachers believed that the integration would help the learners both learn the language and raise environmental awareness as it provided opportunities to read related articles or write texts, "thereby developing their writing skills" (PR12).

The non-linguistic aim of raising learner awareness of climate change and technology was shared by 85 percent of the instructors, 65 percent of the preservice teachers and 27 percent of the learners. A further reason for integration was the relevance of the content as a topic present in public discourse, stated by 36 percent of the instructors, 18 percent of the preservice teachers and eight percent of the learners. Only 19 percent of the preservice teachers regarded the content explicitly conducive to facilitating language learning, while none of the instructors or learners stated this affordance. This is expected from the learners as a pedagogical view cannot be taken for granted within this group, while it may come as a surprise when the instructors' responses are reviewed.

Furthermore, some instructors believed that the content, despite its relevance, conflicted with the school reality in their context:

> In all the language coursebooks, you can find climate change and technology, especially climate change. I try to satisfy my responsibility of raising awareness. However, there has been a problem for many years: the students have difficulties in using the vocabulary/terminology necessary to communicate about environmental pollution, global warming and climate change. Not to mention English, they can't even (without resorting to the internet) deliver a somewhat detailed speech or write a few simple sentences in Turkish. This shows that they have not internalised the content sufficiently. (...) [A student] is told not to throw litter on the ground, and two minutes later you see him/her doing it. (I4)

This response pinpoints the linguistic challenge the content poses and the potential conflict with mindsets within the society. Additionally, the need to gain some science knowledge and develop interest was mentioned by respondents. A preservice teacher remarked that she did not consider integrating the content in her future teaching because "for several teachers (including me) it is difficult to develop instructional ideas about this field of content" (PR19). Another participant in the interview showed no personal interest: "I would never incorporate it in my future teaching because I don't have any idea about it" (PR3).

Ways of Integrating Climate Change and Technology into Language Teaching

For obvious reasons, participants' expertise on how to integrate the content of climate change and technology into language teaching depends on their familiarity with the topic based on experience. Seven instructors reported that they integrated it somehow (one in an unsystematic way), while four stated that they did not. Twenty-nine of the preservice teachers and 24 learners reported having no experience with the content in previous instructions received, while only three and two, respectively, stated that they had.

Four instructors reported that the content was covered in coursebooks, and three instructors stated that they used extra materials. One of them explained that she had designed a poster project on deforestation; some students volunteered to additionally prepare artefacts manufactured with 3-D printers or organised tree planting events. In the survey and interviews, the preservice teachers and learners reflected on how to use technology to address climate change. A number of participants thought that digitalising language teaching through videos, simulations or social media would allow integrating global climate into language teaching and learning. In their responses, technology appeared as a means to make climate change and other global issues accessible as content, but there were no considerations on how to actually convert the content into the subject of a lesson or on the potential benefit of educational technology to engage in remote or hybrid teaching, thereby reducing greenhouse gas emissions.

Discussion and Conclusion

Reporting on a small-scale study, this chapter considers the views of instructors, preservice teachers and language learners on the integration of climate change and technology into foreign language teaching. While the majority of participants considered the integration necessary, their ideas on how to realise such an integration remained vague. The results suggest that the participants did not possess thorough knowledge sufficient to explore climate change and how technology could address this danger. Their unfamiliarity, if not lack of expertise, in science-related content can reasonably be explained by the fact that they were language professionals, students and learners. However, it is crucial to consider different perspectives and experiences when interpreting varying perceptions on the need to integrate climate change and green technology into foreign language teaching. Stated reservations or even resistance, lack of eco-pedagogical and eco-linguistic teaching/learning experiences are likely to place constraints on a desirable integration of climate change and green technology into language teaching, while open-mindedness and corresponding teaching/learning experiences serve as affordances. This result highlights the need to promote foreign language teaching that is enriched through interdisciplinary and cross-curricular content disseminating the concepts of ecology and eco-linguistics (Faramarzi and Janfeshan, 2021; Stibbe, 2015).

The study results also suggest that the Turkish context is characterised by only moderate awareness of environmental issues (Akbana and Yavuz, 2020; Arıkan, 2009; Gürsoy, 2010; Gürsoy and Sağlam, 2011). Participants reported that they found it difficult to unite climate change, education and technology because of conflicting learner behaviour. That said, the participants of this study seized the opportunity to reflect on climate change, technology and education, and how a match between them can be established in language classes.

With an awareness of the restricted local research context and the limited number of participants, it should be pointed out that this chapter serves as a first step to shed light on potential eco-pedagogical and eco-linguistic approaches in foreign language teaching. Studies on designing professional development activities along with the dissemination of good practices of integrating climate change and technology will help develop an area that promises to contribute to both learners' language aquisition and environmental awareness.

References

Akbana, Y.E. and Yavuz, A. (2020). Global issues in EFL teaching: EFL lecturers' voices at a state university. *Kahramanmaraş Sütçü İmam University Journal of Education, 2*(1), 83–102.

Akbana, Y.E. and Yavuz, A. (2022). Global issues in a series of EFL textbooks and implications for end-users to promote peace education through teaching English. *Journal of Peace Education, 19*(3), 373–396. https://doi.org/10.1080/17400201.2022.2140403

Amjad, S., e Zia, A.B. and Masood, M. (2022). Climate change: Students' perspective on eco-linguistic elements in secondary English textbooks. *Human Nature Journal of Social Sciences*, *3*(4), 479–490.

Anderson, A. (2012). Climate change education for mitigation and adaptation. *Journal of Education for Sustainable Development, 6*(2), 191–206. https://doi.org/10.1177/0973408212475199

Arıkan, A. (2009). Environmental peace education in foreign language learners' English grammar lessons. *Journal of Peace Education, 6*(1), 87–99. https://doi.org/10.1080/17400200802655064

Arıkan, G. and Günay, D. (2021). Public attitudes towards climate change: A cross-country analysis. *The British Journal of Politics and International Relations*, *23*(1), 158–174. https://doi.org/10.1177/1369148120951013

Bayraktar Balkır, N. (2021). Uncovering EFL learners' perspectives on a course integrating global issues and language learning. *Novitas-ROYAL, 15*(1), 118–133.

Boon, H. (2014). Teachers and the communication of climate change science: a critical partnership in Australia. *Procedia-Social and Behavioral Sciences*, *116*, 1006–1010. https://doi.org/10.1016/j.sbspro.2014.01.336

Boon, H.J. (2010). Climate change? When or where? *Australian Educational Researcher, 36*(3), 43–65. https://doi.org/10.1007/BF03216905

Chopra, R., Joshi, A., Nagarajan, A., Fomproix, N. and Shashidhara, L. S. (2019). Climate change education across the curriculum. pp. 53–69. *In*: W.L. Filho and S.L. Hemstock (eds.). *Climate Change and the Role of Education.* Springer.

Dahl, T. (2011). Communicating about climate change: Linguistic analysis of climate texts. *Journal of Professional Communication, 26*, 69-75.

Zia, A.B., Sundus, A. and Bhatti, Z.I. (2023). Exploration of ecopedagogical and ecolinguistics elements in secondary level English language textbooks from teacher's perspective. *Journal of Positive School Psychology, 7*(3), 1383–1391.

Faramarzi, A. and Janfeshan, K. (2021). An investigation into ecolinguistics representation in Iranian high school English textbooks. *Ars Artium: An International Refereed Research Journal of English Studies and Culture*, *9*(10), 70–85.

Goulah, J. (2017). Climate change and TESOL: Language, literacies, and the creation of eco-ethical consciousness. *TESOL Quarterly, 51*(1), 90–114. https://doi.org/10.1002/tesq.277

Gray, J. (2016). ELT materials. Claims, critiques and controversies. pp. 86–113. *In*: G. Hall (ed.). *The Routledge Handbook of English Language Teaching*. Routledge.

Gürsoy, E. (2010). Implementing environmental education to foreign language teaching to young learners. *Educational Research, 1*(8), 232–238.

Gürsoy, E. and Sağlam, G.T. (2011). ELT teacher trainees' attitudes towards environmental education and their tendency to use it in the language classroom. *Journal of International Education Research, 7*(4), 47–52. https://doi.org/10.19030/jier.v7i4.6046

Inayati, N., Adityo, A. and Hima, A.N. (2016). Environmental awareness raising contents in k13 English textbooks published by the Indonesian ministry of education. *Prosiding Seminar Nasional II Tahun [Tahun Proceedings of the National Seminar]* (pp. 302–312).

Jodoin, J.J. and Singer, J. (2018). An analysis of environmental content found in English-language textbooks in Japanese higher education using a corpus. *International Journal of Social Sustainability in Economic, Social & Cultural Context, 14*(4), 39–55. https://doi.org/10.18848/2325-1115/CGP/v14i04/39-55

McGrath, I. (2013). *Teaching materials and the role of EFL/ESL teachers.* Bloomsbury.

Mohammadnia, Z. and Moghadam, F.D. (2019). Textbooks as resources for education for sustainable development: A content analysis. *Journal of Teacher Education for Sustainability*, *21*(1), 103–114. https://doi.org/10.2478/jtes-2019-0008

Ramadhan, S., Sukma, E. and Indriyani, V. (2019). Environmental education and disaster mitigation through language learning. *The 1st International Conference on Environmental Sciences (ICES2018) IOP 314*(1), 1–9. https://doi.org/10.1088/1755-1315/314/1/012054

Saldana, J. (2009). *The Coding Manual for Qualitative Researchers.* Sage.

Stibbe, A. (2004). Environmental education across cultures: Beyond the discourse of shallow environmentalism. *Language and Intercultural Communication, 4*(4), 242–260. https://doi.org/10.1080/14708470408668875

Stibbe, A. (2015). Ecolinguistics: Language, ecology and the stories we live by. Routledge.

Tenridinanti, T.B., Inderawati, R. and Mirizon, S. (2021). Report texts on climate change: Elaboration of students' needs analysis. *Prosiding Seminar Nasional Bahasa Dan Sastra (Senabatra) 1*(1), 78–86.

UNESCO (United Nations Educational, Scientific, and Cultural Organization) (2010). The UNESCO climate change initiative: Climate change education for sustainable development. UNESCO. https://unesdoc.unesco.org/images/0019/001901/190101E.pdf.

Yasukawa, K. (2023). Teaching about climate change: Possibilities and challenges in Australian adult literacy programs. *Journal of Adolescent & Adult Literacy, 66*(4), 218–228. https://doi.org/10.1002/jaal.1267

Yılmaz, F. and Güleryüz Adamhasan, B. (2022). Representation of global issues in EFL textbooks. pp. 185–193. *In*: Y. Özkan, E.B. Deniz, A. Dinçer, B.B. Harmandar, D. Sucak, T. Ekiz and M. Kara (eds.). *First International Language-for-all Conference. Book of Proceedings*. School of Foreign Languages, Çukurova University.

Chapter 8

Enacting Climate Education in Teacher Preparation and Professional Learning

Martha C. Monroe,[1,*] *Kathryn Riley*[2] and *Peta J. White*[3]

Students are changing faster than teacher preparation and professional learning

The recent book by Andreas Malm (2021) and movie of the same name, *How to Blow Up a Pipeline*, present a compelling case demonstrating the frustration that many young people experience about climate change (Boulianne et al., 2020). Greenhouse gas emissions have increased global surface temperatures by 1.1°C over the average temperatures in the late 1800s (IPCC, 2022). While there is some variation by region, income, and level of development, emissions are still increasing, and communities who contribute the least to climate change are often those most impacted by it (IPCC, 2022). Yet the warnings of impacted ecosystems, international gatherings and proclamations, and municipal initiatives to achieve a net zero carbon future, seem to have little impact on the lives of most people, who continue to drive to work, air condition

[1] University of Florida, Gainesville, Florida, USA.

[2] University of Manitoba, Winnipeg, Canada.
Email: Kathryn.riley@umanitoba.ca

[3] Deakin University, Melbourne, Australia.
Email: peta.white@deakin.edu.au

* Corresponding author: mcmonroe@ufl.edu

their homes, and eat as high on the food chain as they always have (Anderson, 2012; Lee, 2021).

With little evidence of real change from policy makers and institutions, many young people are anxious and fearful for their futures (Bright and Eames, 2022; Hickman et al., 2021) and some are frustrated at being silenced in the face of obvious injustices. They bring these frustrations, fears, and big emotions into the classroom. Many of our students are now more politically savvy and active, concerned, and demanding to be heard than with past generations. The international *School Strike 4 Climate* and *Futures Fridays* movements (See in Australia: https://www.schoolstrike4climate.com/; Verlie and Flynn, 2022) demonstrate the student-led and intergenerational concerns about the lack of climate change education and political mandates towards sustainable (future focussed) decisions and fossil fuel reductions. Young people are demanding their voices be heard and their futures protected (White et al., 2021). They organise political action that is targeted, focussed, and impactful. They are enacting civil disobedience in a system that doesn't typically enable young people to have a voice in major decisions about their futures.

School and teacher responses to our changed reality and politically capable young people are varied. Many teachers feel they lack the skills to address inevitable and understandable student eco-anxiety (Finnegan, 2022), and feel unprepared to respond to student questions and comments that could both trigger despair or be channelled into hopeful action (Ojala, 2023a, 2023b). Preparing teachers to enact climate change education requires that they understand climate change, the local and global impacts of the multiple climate crises, and their students' reactions. Attention should be given to the diversity of perspectives engaged in building sustainable futures for all. Opportunities to redevelop pre-service teacher preparation and professional learning are essential, and some exemplars are presented in this chapter.

The way that school curricula enable climate change education varies across jurisdictions and nations (Dawson et al., 2022), which may require that schools and teachers make their own decisions about what to include in the classroom. Teachers must be supported and enabled to both meaningfully engage with students' emotional reactions to climate change, and to foster skills that will empower students to play an active role in addressing climate challenges.

Support to Enable Change in Education

Two recent documents provide suggestions to educators as they address climate literacy in young people. The first is the Organisation for Economic and Cooperative Development's (OECD) Programme for International Student Assessment (PISA) 2025 *Science Framework* (https://pisa-framework.oecd.org/science-2025/). The PISA tests science, reading, and mathematics competencies of 15-year-old students on a three-year rotation. The assessment is based on a framework developed by a group of content and education experts. The 2025 Science Framework defines *Agency in the Anthropocene* and provides an additional support document (White et al., 2023) that details competencies about understanding human impacts on Earth systems, the need to respect culturally diverse ways of knowing, and consideration for all species and beings. The OECD has ensured that these documents are accessible so that education policy makers can use them when designing or reforming their curricula. These documents provide guidance to restructure curricula to better prepare 15-year-old students to succeed in the PISA assessment and as responsible decision makers and engaged citizenry. Many countries desire success with PISA results. Working with the Agency in the Anthropocene document provides insights and suggestions regarding ways to embed climate change education across disciplines (White et al., 2023).

Concurrently, the North American Association for Environmental Education (NAAEE) is developing the second key document: *Educating for Climate Action and Justice: Guidelines for Excellence*, to complement their existing series of guideline documents (NAAEE, 2023). The guidelines for climate education provide suggestions for educators, in both formal and nonformal education programs, to enable them to build knowledge, attitudes, skills, and commitments among their learners. Unlike other documents about climate literacy, these guidelines seek to empower learners to engage in climate justice and action through five key characteristics: (1) Collaborative, Welcoming, and Responsive Learning Environment; (2) Knowledge and Skills to Foster Climate Action, (3) Climate Emotions, (4) Locally Focused and Community Driven, and (5) Civic Engagement for Climate Action. Schools and teachers can use the guidelines to ensure that learning attends to climate change education in meaningful ways. The guidelines support educators preparing their students to be engaged in local action to address current injustices and strengthen skills to manage inevitable challenges while developing communities that sustain this kind of effort.

Considered together, these documents focus on integrated, interdisciplinary, and relational learning that enables young people to work individually and collectively towards solutions to problems in classrooms, schools, and their communities. Addressing these goals requires new perspectives in teacher professional learning, that include conceptual understanding, pedagogical tools, and support systems that empower teachers to engage learners in action projects. Teacher preparation and professional learning will be essential to institutionalise these new practices and culture. These two documents operate at different levels (the OECD assessment offers insight for national/state curriculum policy, whereas the NAAEE Guidelines can be used at the program level with teachers, non-formal leaders, or even evaluators), but when enacted together both documents provide powerful rationale and vision to change the ways schools and teachers offer contemporary learning to young people. The uncertain nature of our futures is confusing. One aspect of our future is certain, however: the world will not be the same as we know it today.

Addressing Key Challenges for Teachers Enacting Climate Change Education

Numerous studies report what teachers already know about climate change, with common themes occurring the world over, despite differences in education systems, government support, and environment. Teachers' base knowledge of climate science generally reflects common beliefs, including misconceptions (Boon, 2010; Wise, 2010; Wojcik et al., 2014). As public knowledge of climatic changes has increasingly aligned with the scientific consensus over the last decade, so too have teachers' perspectives, with more believing that humans have caused recent changes to the planet's climate (Wise, 2010; Plutzer et al., 2016). Those who disagree may not be misinformed, but rather be influenced by political ideology, which is often a stronger predictor of climate beliefs than content knowledge (Kahan, 2015; Kunkle and Monroe, 2019; Plutzer et al., 2016). In many countries, teachers are more than willing to teach climate change and there is strong public support for schools addressing the issue (Dawson et al., 2022; Kamenetz, 2019; Worth, 2021). Despite this, teachers often perceive a lack of political and curricular support (Worth, 2021), a lack of time and resources

(Ennes et al., 2021), and believe that they need additional professional development (Rodler and Renbarger, 2023) in order to deliver effective climate education. These barriers need to be addressed, indeed in several countries work is underway to create federal policies that reflect the United Nations' Sustainable Development Goal #13; taking action on climate change through education.

Climate change education must be included in both pre-service and in-service professional learning for teachers to be able to explore it with their students (White, et al., 2024). Recommendations are available for what such learning might include, based on reviews of what constitutes effective climate education (Cantell et al., 2019; Monroe et al., 2019). This literature has laid the foundation for both the OECD PISA 2025 *Science Framework—Agency in the Anthropocene* and NAAEE's guidelines for climate education. Both documents speak to three goals of teacher preparation and professional development: shifting towards trans-disciplinary educational practice; addressing emotions; and empowering actions. We will now examine these goals with examples from each document.

Moving to Transdisciplinary Curriculum

Effective climate change education enables a focus on disciplinary learning that intersects and reinforces learning through multiple disciplines. Disciplinary knowledge would include climate science (including an understanding of the models that reflect measurable changes in Earth's systems, the areas in which there is uncertainty, the reasons for climate change, and the likely impacts of these changes on the environment) separate from the many ways climate change is impacting social systems and the costs and benefits of various solutions to mitigate and adapt to climate change. Effective climate education spans and links these disciplines, allowing students to build a transdisciplinary understanding of a complex topic (Raphael et al., 2021). Teachers not only need to be informed and understand the current scientific consensus but must also be able to recognise and address common misconceptions, prevalent across social media, that their learners might bring into the classroom (Osborne and Pimentel, 2022). Similarly, educators who conduct teacher professional learning or initial teacher education must also be able to address preconceived notions, particularly those that are socially constructed and lack factual grounding. Acknowledging impacts across ecosystems and communities requires sensitivity, a sense of climate justice, and a willingness to consider strategies that reduce disparity and social injustices. Educators should also be aware of, acknowledge and convey multiple ways of knowing about climate change (these are epistemological perspectives often related to cultural connections) and the variety of perspectives on how society can mitigate and/or adapt. Given that some perspectives are strongly associated with political positions, and may be controversial, teachers must also develop skills to navigate challenging discussions and consider multiple stakeholder perspectives. Climate change education is best when considered in transdisciplinary ways through multiple and diverse lenses.

Climate change education may include ecological, political, ethical, sociocultural, spiritual, and material forces. It considers and enables understanding from diverse, multiple, intergenerational perspectives, and an appreciation of complex systems. It encourages socio-ecological challenges to be considered holistically, from how they are produced to the impacts they create, and the ways in which societies both perpetuate and adapt to them. Subject complexity is important, as studying isolated examples or systems likely results in false or misleading conclusions.

An analysis of OECD PISA data from 2018 suggests that youth understand environmental issues and care about them, but do not know how to contribute to their resolution (OECD, 2022). The PISA 2025 *Science Framework* stretches the traditional boundaries of science education to include ways communities can mitigate and adapt to climate change, systems-thinking, decision-making and action-taking skills, and includes diverse perspectives on both problems and solutions to socio-ecological crises (White et al. 2023). This could empower science educators to dabble in the social sciences of policy and economics or teach courses that include strategies of addressing climate change. Similarly, NAAEE's climate education guidelines list systems-thinking as an essential skill, linking climate science with climate solutions and climate justice.

Addressing Emotions

Climate change generates a variety of emotions, from concern to anxiety, guilt to despair, and interest to anger (Bright and Eames, 2022), in both teachers and students. Recognising and acknowledging these feelings can enable individuals to process them and consider their rights and responsibilities. Often, this done best within a community and in collaboration with colleagues or friends. Recognizing success stories, champions, and strategies that improve community well-being can build hope for the future. Effective climate education prepares instructors to adapt learning strategies to engage with emotions and to effectively address them.

One of the five key characteristics in NAAEE's climate education guidelines speaks to the importance of addressing emotional responses, providing time for reflection, and legitimising mental health. Some degree of concern, guilt, anger, and anxiety is expected and may galvanise appropriate action. Educators should be aware of local mental health resources that students can access and also convey positive outlooks of potential solutions and work being done towards achieving a sustainable (Chawla, 2020; Ratinen and Uusiautti, 2020).

Enabling Action

Empower learners to address climate change solutions, teachers must provide a meaningful vision of possibility through examples of successful action and foster the skills their students need to take part in such actions. This should include a variety of skills and strategies that can promote action, such as: systems, critical, and creative thinking; recognizing the NGOs, agencies, and decision makers who have relevant power; working together to listen and lead; seeking and partnering with resource people in the local community; advocating with accuracy, persuasiveness and passion; and planning, evaluating, and conducting action projects. Several of these strategies have been part of active civic education and environmental education programmes for decades. The teacher's role is to understand and appreciate their learners, to know when and how to guide them towards success, and to allow their students to lead the process. Teacher preparation programs can develop pre-service teachers' sense of civic responsibility, awareness of community agencies and organisations that support citizen actions, and the appropriate channels through which residents may interact with municipal governance (Monroe et al., 2016). Climate change education is inherently political, on many levels.

Both the OECD PISA 2025 *Science Framework* and *Agency in the Anthropocene* (White et al, 2023) and NAAEE's climate education guidelines recognise that while conducting action projects may be the best way to gain skills and understanding around civic engagement (Monroe et al. 2019), many teachers may not have time to devote to a long-term project. Developing cross-curricula learning opportunities and guided inquiry projects are skills to be learned. School leaders can assist in creating these learning opportunities. Examples include inviting local speakers to provide an analysis of complex socio-ecological challenges, involving youth in citizen science programs and service-learning opportunities, and initiating school-based projects. After-school clubs and other non-formal education programs can also play a crucial role (Monroe et al. 2023; Ardoin, et al. 2023; Schusler and Krasny, 2008).

The Need for Ongoing Teacher Education and Development of High-quality Teaching Resources: The Application of Technology

For many educators, transdisciplinary approaches, addressing students' emotional needs, and empowering actions might represent a significant shift from traditional teaching practices, and, as such, may go beyond what they have encountered in teacher preparation programs. The need for ongoing professional development and availability of resources is key. The complexity and opportunities that climate change brings to education, necessitates a rethinking of how we support teachers and develop resources. The following three examples focus on teaching practices relevant for teacher preparation, in-service professional development, and resource development and implementation, each with a strong technological focus.

Teacher Preparation Courses Adapting to Climate Change Education: the *Digiexplanation*

Pre-service courses and units, in-service workshops and coursework, and instructional materials can be designed to prepare and support educators to think and teach differently and meet the challenges that climate change presents. For example, a bachelor's level teacher education course at the University of Manitoba in Winnipeg, Canada has the following objectives:

a) Orientation to climate change education (CCE) and locating CCE in Manitoba K-12 curriculum.
b) Systems thinking and Socio-scientific Inquiry-based Learning (SSIBL).
c) Climate change and sustainability.
d) Philosophical conceptualising of nature, land, and place, including an examination of colonial systems.

e) Indigenous land-based learning through counter-oppressive pedagogies and assemblage thinking.
f) Practical teaching and learning strategies in, and for, CCE.
g) Transdisciplinary approaches to curriculum.
h) Active citizenship for climate justice and communication and persuasion through the media.
i) Emotional inquiry and learning to live with climate change.

In this course, pre-service teachers came from a wide variety of teaching disciplines, increasing the importance of contextualising CCE to promote meaningful and relevant content that could be later adopted in their professional practices. Thus, this course provided many opportunities to grapple with transdisciplinary approaches to curriculum. A major assessment task in the course involved the artistic creation of a *digiexplanation*. Using digital animation (stock and personal video and images) and video-maker technologies), the task invited learners to clarify their understanding of climate change by explaining and providing examples of natural phenomena *and* social contexts relevant to their disciplinary specializations in the Bachelor of Education program. Embodying transdisciplinary approaches to curriculum, learners began creating their *digiexplanations* starting with a given socio-ecological problem. They then sought to make sense of this problem by making connections between and across knowledges and practices from diverse curricular disciplines (Riley and Proctor, 2022; Riley and White, 2019; Tinnell, 2012).

Problem-based inquiry, the creation of the *digiexplanations* provided learners with the necessary conditions to expand across natural/social and science/art boundaries, with an awareness of their own (micro) politics, and ethno-ancestral, religious, socioeconomic, and linguistic positionalities and backgrounds. While not an exhaustive list, some examples of the diverse socio-ecological problems that learners explored include: figuration of the Plains-bison as a symbol of colonial land practices that have removed Indigenous peoples; prairie landscapes as one of the most threatened and endangered ecosystems in the world through the conversion of diverse grasslands to agricultural crops and flood-plain monocultures (Hisey et al., 2022; Mueller et al., 2021; Samson et al., 2004); implications of climate change on Australia's dying coral reefs (Ramis et al., 2012); heavy use of fertilizers in agricultural practices across the Canadian prairies causing toxic blue-green algae in Lake Winnipeg and serious health and safety concerns for lake recreation

(Sommerfeld, 2012); the Fukushima nuclear accident in Japan and its effects on ecological systems of the Pacific eastern seaboard due to wind and ocean currents spreading toxic fallout waste (Murakami and Reuters, 2023); Indigenous articulations of waters of the earth and waters of the body as the same, and thus, colonial and capitalist exploitation of the planet being an exploitation of the body (Yazzie and Baldy, 2018); explorations of Robert Smithson's *spiral jetty* as a way to articulate how land art can teach us about climate change and socio-ecological crises (Ballard and Linden, 2019).

The *digiexplanations* provided a platform for learners to experiment with the complexity of interrelated and interconnected socio-ecological problems that contribute to climate change. Crucially, however, the *digiexplanations* also invited learners to engage in emotional inquiry in response to living with impoverished socio-ecological systems in these times of the Anthropocene. As Randall (2009) claimed, climate change discourses present two parallel narratives. The first is about the problems of climate change and the second is about solutions to climate change. A major feature in narratives about the problems of climate change is the idea of loss somewhere out in the future, and in other remote parts of the world usually detached from Western audiences. Within narratives about solutions, loss does not seem to bare the same prominence. For Randall, these parallel narratives are divided as the result of a defensive process to protect public from having to mourn and grieve losses associated with climate change. Yet, due to global warming, as natural disasters become more frequent and more extreme, many people are experiencing chronic fear of an environmental cataclysm. This fear comes from observing the seemingly irrevocable impact of climate change and the associated concern for one's own future and that of future generations. For instance, new climate change education research suggests that young people are experiencing significant eco-anxiety and climate grief due to internalising the grave environmental problems the planet is facing, and as a response to the loss of cherished species, ecosystems, and landscapes (Bright and Eames, 2022; Cunsolo and Ellis, 2018; Verlie and Flynn, 2022). Thus, to enact effective and affirmative change, it is important to learn how to cope with eco-anxiety, climate grief, and other emotional responses to socio-ecological injustice and threats, thereby ensuring personal wellness over time.

It was in this spirit that the *digiexplanations* inspired ongoing class discussions to purposefully explore a range of emotional responses to local and global socio-ecological problems. Problems that were exacerbated by climate change and further contributed to its ongoing and pervasive effects. A deliberate focus on emotional inquiry in this course served to initiate care ethics and practical action, within calls for justice in the face of climate change steeped in racist, classist, and gendered processes of exploitation and domination (McKenzie et al., 2023), social and ecological precarity, and threats unevenly distributed across the planet (Monroe et al., 2019; Stevenson et al., 2017). The goal was for pre-service teachers to adopt climate activism along diverse teaching pathways in their own professional practices; pathways that disrupt the limitations of contemporary education systems to generate sustainable futurity and ethical horizons of hope and professional practices that are situated, emplaced, and contextualised within the lived stories of the educational community.

Teacher Education Course Climate Change Education Opportunities with Species Identification Apps

A graduate certificate qualification designed for 'out-of-field' teachers who are currently teaching in fields in which they have no undergraduate qualification (White and Peters, 2021), is one strategy for providing in-depth support for qualified educators. Undertaking a qualification requires a significant commitment, as opposed to completing a professional development seminar. In a qualification, assessment and evaluation is paramount. This course, designed for fully qualified teachers, focussed on learning content through appropriate pedagogies. The course was comprised of four units, one being focussed on biology. Attention was given to considering contemporary biology, which included listening to scientists discuss their research regarding local socio-ecological challenges. The species app was useful to tune participants into thinking about different organisms, their needs, and how they deal with climate impacts (White and Peters, 2021).

Realising experiences of impact is important in understanding the reality of climate change. Climate denial, shrouded in misinformation (Osborne and Pimentel, 2022), is enabled by polarised arguments in the media that are framed as a binary (yes and no, right and wrong). Students require complex skills to judge accuracy and reliability of information. An additional strategy to scaffold and support students in understanding the impacts of climate change, is to consider species located in backyards and local parks. This is possible through a range of species identification apps. For example, digital tools (species identification apps) such as "*Seek*" by iNaturalist enables the development of online communities of people interested in mapping and monitoring species in their backyards. Task designs could vary, relating to learner interests directly as well as curriculum content. Creating choices about how the learner might apply the app is a useful strategy to ensure that the activity feels engaging and of interest. Choices included:

- Identify two unknown species. Draw each and identify body parts (you may need to use additional information to determine the names of some body parts). Based on your observations, describe the habitat requirements for each individual. Describe two structures for each individual that enables this species to survive in their environment; these may be adaptations to their ecosystem niche.
- Photograph an organism that you notice in your local area. Search for the scientific name (genus and species). Search the app for a similar species, based on any characteristic. Consider the characteristic as a feature for survival and describe the advantages and potential disadvantages of the characteristic.
- Find a local natural area that has local (endemic) species and introduced species. Use the app to find out which species are introduced. Describe why this species thrives in this area. Describe the characteristics that ensure its survival. Discuss the similarities and differences between these characteristics and those of local species.
- Locate a pollinator species and identify it using the App. Draw it highlighting the key features of the species (especially how it is a pollinator). Represent the food chain of that species. Consider what might happen if one species of that food chain was to disappear.

These activities rely on the use of the app and connect the learning to local environments and species, which can be an effective way to demonstrate relevance for many students. The app can be used initially and then, if interest is piqued, additional digital tools such as the *Atlas of Living Australia* (https://www.ala.org.au/) can facilitate projects that explore species populations and distributions. Data collection from citizens (students) is enabled and tools for data analysis are available. These tools can be used in multiple ways. In educational settings, the tools provide educational supports, enabling teachers to design their own explorations. These flexible tools can ensure that activity design can be adapted to tap into complex socio-ecological challenges such as the management of introduced species, species population and distribution changes, recovery after bushfires, and many other climate impacts.

Instructional Resources and Professional Learning Build Teacher Efficacy in Climate Change

Opportunities to support educators teaching about climate include the development of instructional material that offers trustworthy background information, suggestions for engaging activities, and supplemental resources such as slide presentations, videos, and worksheets. Technology, in the form of a website, can provide additional value by bringing these resources together and providing easy access to educators. The development of *Southeastern Forests and Climate Change* (Monroe and Oxarart, 2015) is a good example of a regional climate education resource that uses technology to enhance student and teacher understanding of climate issues and potential solutions. In addition, several activities engage students in using technologies, such as basic tools for measuring tree diameter and height to calculate sequestered carbon, as well as computer-driven models for climate projections.

This instructional resource was made available through educator professional learning workshops and a website. Three workshops were led by the design team to train facilitators; 86 workshops were led by facilitators across the region reaching a total of 1411 educators over two years. Several strategies were essential to strengthen teachers' efficacy in teaching climate change in secondary science classrooms.The material was designed to encourage transdisciplinary, hope-filled, and action-oriented education. For example, an initial needs assessment asked science teachers if they were willing to teach about life cycle analysis in the context of decisions youth could make to reduce their carbon footprint. While this information is typically introduced in an economics or geography course, biology and environmental science teachers who responded were enthusiastic about addressing this type of solution. Three of the 14 activities focused on life cycle analysis to estimate the atmospheric carbon produced by equivalent products. Other activities introduced the carbon cycle, genetic characteristics, and then applied this background science to forest management challenges. Through these and a series of systems-thinking activities that accompanied each climate activity, teachers were encouraged to link various sciences together (biology, chemistry, meteorology) and connect them to social science topics. Culminating activities guided students to use their knowledge to address transdisciplinary problems.

Several activities directed educators to use interactive models and data from websites provided by federal agencies. While external sites may not be available in the long term, some are likely to be updated and offer current data and explanations. Introducing students to strategies to manipulate decision-making models enables them to become more comfortable with innovative technology.

To ensure students recognise that people are working on solutions to climate change, activities featured stories of people managing forests to address climate changes, researchers working to improve forest growth in a changed climate, and industry using long-lasting wood products and planting replacement forests to sequester even more carbon, thus reducing atmospheric carbon. Activities provided enough information for students to recognise why and how these actions contributed to reduced atmospheric carbon and generated hope.

To support students in considering actions they could take to make a difference, the life cycle analysis activities used examples of products young people purchase and featured career opportunities.

Based on pre and post data collected at workshops, 55% of the participants had not previously taught about climate change, 67% were very likely to use the module in their classrooms, and 76% would recommend the material to colleagues (Monroe and Oxarart, 2019). Pre and post scores of efficacy for teaching about climate change showed a significant increase among all workshop participants, both those who had previously taught about climate and those who were new to climate education (Li et al., 2019). This suggests the materials themselves, the common denominator across all workshops, might be largely responsible for the increase in efficacy.

Ongoing Education for Schools, Teachers, and Students Supports Innovation

Climate change is heralding change in every sector of society, and education reform is required to ensure young people have the support necessary to understand their potential futures. This chapter addressed these reform challenges with suggestions for ways education can be adapted to meet the needs of young people. Educational reform will require reconceptualisation of teacher education programs to ensure that teachers are prepared and receive ongoing support through professional learning and well-designed instructional resources. Reflecting on what we are doing and why we are doing it enables us to continue our own process of learning and improving our practice.

In this chapter several strategies for teacher education that focus on climate change education have been presented. The OECD PISA 2025 *Science Framework* (https://pisa-framework.oecd.org/science-2025/) and *Agency in the Anthropocene* support document (White et al., 2023) and NAAEE's guidelines for climate change education provide insights, clarity, and strategies and support to effectively address climate change. These documents are calls to action at the curriculum policy and classroom level. They can be used by administrators and school leaders, teacher educators, and teachers to bring about necessary change and educational reform.

Technology is a key tool in climate change education. As teachers prepare young people to consider uncertain futures with careers not yet determined (See Tytler et al., 2019: https://100jobsofthefuture.com/), they will likely require innovative tools and support. Understanding how managers, planners, and educators currently use technology to explore and explain the world and mimicking their practice can improve skills. Summaries of existing effective climate change education and indicators of ideal climate education suggest three directions that will likely create change in current teacher preparation programs: shifting to transdisciplinary education, addressing emotions, and enabling action. We have described these and offered examples of how we are working toward educational programs that strengthen teachers' capacity to empower their learners.

References

Anderson, K. 2012. The inconvenient truth of carbon offsets. *Nature*, 484, 7. https://www.nature.com/articles/484007a

Ardoin, N.M., Bowers, A.W. and Gaillard, E. (2023). A systematic mixed studies review of civic engagement outcomes in environmental education. Environmental Education Research, 29(1), 1-26. https://doi.org/10.1080/13504622.2022.2135688

Atlas of Living Australia (2024). Retrieved from https://www.ala.org.au/Ballard, S. and Linden, L. (2019). Spiral Jetty, geoaesthetics, and art: Writing the Anthropocene. *The Anthropocene Review*, *6*(1–2), 142–161. https://doi.org/10.1177/2053019619839443

Boon, H.J. (2010). Climate change? Who knows? A comparison of secondary students and pre-service teachers. *Australian Journal of Teacher Education*, *35*(1), 104–120. https://researchonline.jcu.edu.au/8231/1/AJTE_Boon_Feb_2010.pdf

Boulianne, S., Lalancette, M., Ilkiw, D. (2020). "School Strike 4 Climate": Social media and the international youth protest on climate change. *Media and Communication*, 8:2, 208–218. https://www.ssoar.info/ssoar/bitstream/handle/document/67824/

ssoar-mediacomm-2020-2-boulianne_et_al-School_Strike_4_Climate_Social.pdf?sequence=1

Bright, M.L. and Eames, C. (2022). From apathy through anxiety to action: Emotions as motivators for youth climate strike leaders. Australian Journal of Environmental Education, *38*(1), 13–25. https://doi.org/10.1017/aee.2021.22

Cantell, H., Tolppanen, S., Aarnio-Linnanvuori, E. and Lehtonen, A. (2019). Bicycle model on climate change education: Presenting and evaluating a model. Environmental Education Research, *25*(5), 717–731. https://doi.org/10.1080/13504622.2019.1570487

Chawla, L. (2020). Childhood nature connection and constructive hope: A review of research on connecting with nature and coping with environmental loss. *People and Nature*, *2*(3), 619–642. doi:10.1002/pan3.10128

Cunsolo, A. and Ellis, N.R. (2018). Ecological grief as a mental health response to climate change-related loss. Nature Climate Change, *8*(4), 275–281. https://uwosh.edu/sirt/wp-content/uploads/sites/86/2020/04/Cunsolo-and-Ellis-2018.pdf

Dawson, V., Eilam, E., Widdop-Quinton, H., Putri, G. A. P. E., White, P., Subiantoro, A. W., Tolppanen, S. Gokpinar, T., Goldman D. and Ben-Zvi Assaraf, O. (2022). A Cross-country comparison of climate change in middle school science and geography curricula. International Journal of Science Education. 44(9), 1379–1398. https://doi.org/10.1080/09500693.2022.2078011

Ennes, M., Lawson, D.F., Stevenson, K.T., Peterson, M.N. and Jones, M.G. (2021). It's about time: perceived barriers to in-service teacher climate change professional development. *Environmental Education Research*, *27*(5), 762–778.

Ergler, C.R., Kearns, R., Witten, K. and Porter, G. (2016). Digital methodologies and practices in children's geographies. *Children's Geographies, 14*(2), 129–140. https://doi.org/10.1080/14733285.2015.1129394

Finnegan, W. (2022). Educating for hope and action competence: A study of secondary school students and teachers in England. *Environmental Education Research*, 1–20.

Hickman, C., Marks, E., Pihkala, P., Clayton, S., Lewandowski, R.E., Mayall, E.E. and Van Susteren, L. (2021). Climate anxiety in children and young people and their beliefs about government responses to climate change: a global survey. *The Lancet Planetary Health*, *5*(12), e863–e873.

Hisey, F., Heppner, M. and Olive, A. (2022). Supporting native grasslands in Canada: Lessons learned and future management of the Prairie Pastures Conservation Area (PPCA) in Saskatchewan. *The Canadian Geographer/Le Géographe Canadien*, *66*(4), 683–695.

IPCC, 2022: Summary for Policymakers. *In*: Climate Change 2022: Mitigation of Climate Change. Contribution of Working Group III to the Sixth Assessment Report of the Intergovernmental Panel on Climate Change [P.R. Shukla, J. Skea, R. Slade, A. Al Khourdajie, R. van Diemen, D. McCollum, M. Pathak, S. Some, P. Vyas, R. Fradera, M. Belkacemi, A. Hasija, G. Lisboa, S. Luz, J. Malley, (eds.)]. Cambridge University Press, Cambridge, UK and New York, NY, USA. doi: 10.1017/9781009157926.001

Kahan, D.M. (2015). Climate-science communication and the measurement problem. *Political Psychology*, *36*, 1–43.

Kamenetz, A. (2019). Most teachers don't teach climate change; 4 in 5 parents wish they did. National Public Radio. Climate Change Earth Day Poll: 4 In 5 Parents Want It In Schools: NPR.

Kunkle, K.A. and Monroe, M.C. (2019). Cultural cognition and climate change education in the US: Why consensus is not enough. *Environmental Education Research, 25*(5), 633–655.

Land, N., Hamm, C., Yazbeck, S.L., Danis, I., Brown, M. and Nelson, N. (2019). Facetiming common worlds: Exchanging digital place stories and crafting pedagogical contact zones. *Children's Geographies, 18*(1), 30–43. https://doi.org/10.1080/14733285.2019.1574339

Lee, M. 2021. *Dangerous Distractions: Canada's carbon emissions and the pathway to net zero.* Ottawa: Canadian Centre for Policy Alternatives.

Li, C.J., Monroe, M.C., Oxarart, A. and Ritchie, T. (2019). Building teachers' self-efficacy in teaching about climate change through educative curriculum and professional development. *Applied Environmental Education & Communication,* DOI: 10.1080/1533015X.2019.1617806

Malm, A. (2021). *How to Blow Up a Pipeline*. Verso Books. ISBN 978-1-83976-025-9.

McKenzie, M., Henderson, J. and Nxumalo, F. (2023). Climate change and educational research: Mapping resistances and futurities. *Research in Education, 0*(0), 1–8. doi: 10.1177/0034523723120307

Monroe, M.C., Ballard, H.L., Oxarart, A., Sturtevant, V.E., Jakes, P.J. and Evans, E.R. (2016). Agencies, educators, communities and wildfire: Partnerships to enhance environmental education for youth. *Environmental Education Research, 22*(8), 1098–1114. https://doi.org/10.1080/13504622.2015.1057555

Monroe, M.C., Eames, C., White, P.J. and Ardoin, N.M. (2023) Education to build agency in the Anthropocene, *The Journal of Environmental Education, 54*(6), 351–354, DOI: 10.1080/00958964.2023.2277209

Monroe, M.C. and Oxarart, A. (eds.) (2015). *Southeastern Forests and Climate Change: A Project Learning Tree Secondary Module*. Gainesville, FL: University of Florida and American Forest Foundation.

Monroe, M.C. and Oxarart, A. (2019). Integrating research and education: Developing instructional materials to convey research concepts. *BioScience* 69(4): 282–291. https://doi.org/10.1093/biosci/biz008

Monroe, M.C., Plate, R.R., Oxarart, A., Bowers, A. and Chaves, W.A. (2019). Identifying effective climate change education strategies: A systematic review of the research. *Environmental Education Research, 25*(6), 791–812. doi: 10.1080/13504622.2017.1360842

Mueller, N.G., Spengler III, R.N., Glenn, A. and Lama, K. (2021). Bison, anthropogenic fire,and the origins of agriculture in eastern North America. The Anthropocene Review, *8*(2), 141–158. https://doi.org/10.1177/20530196209611

Murakami, S. and Reuters, T. (2023, August 24). *Japan releases wastewater from Fukushima nuclear plant into Pacific Ocean amid protests*, CBC news. https://www.cbc.ca/news/world/japan-releases-fukushima-wastewater-1.6945944

NAAEE [North American Association for Environmental Education]. (2023). Guidelines for Excellence in Environmental Education series. https://naaee.org/programs.guidelines-excellence

OECD [Organisation for Economic Cooperation and Development]. (2022). Are students ready to take on environmental challenges? PISA. OECD Publishing. https://doi.org/10.1787/ 8abe655c-en

Ojala, M. (2023a). Climate-change education and critical emotional awareness (CEA): Implications for teacher education. *Educational Philosophy and Theory*, 55:10, 1109-1120, DOI: 10.1080/00131857.2022.2081150

Ojala, M. (2023b). Hope and climate-change engagement from a psychological perspective. *Current Opinion in Psychology*, *49*, 101514.

Osborne, J. and Pimentel, D. (2022). Science, misinformation, and the role of education. *Science, 378*(6617). https://www.science.org/doi/10.1126/science.abq8093

Plutzer, E., McCaffrey, M., Hannah, A.L., Rosenau, J., Berbeco, M. and Reid, A.H. (2016). Climate confusion among US teachers. *Science*, *351*(6274), 664–665.

Raphael, J., White, P.J. and van Cuylenburg, K. (2021). Sparking Learning in Science and Drama: Setting the Scene. *In*: P.J. White, J. Raphael and K. van Cuylenburg (eds.). *Science and Drama: Contemporary and Creative Approaches to Teaching and Learning*. (Chapter 1, pp 1–25). Springer.

Ramis, M. and Prideaux, B. (2012). The importance of visitor perceptions in estimating how climate change will affect future tourist flows to the Great Barrier Reef. pp. 173–188. *In*: M.V. Reddy and K. Wilke (eds.). *Tourism, Climate Change and Sustainability*. Routledge.

Randall, R. (2009). Loss and climate change: The cost of parallel arratives. *Ecopsychology*, 1(3), 118–129. http://www.liebertpub.com/products/product.aspx?pid=300

Ratinen, I. and Uusiautti, S. (2020). Finnish students' knowledge of climate change mitigation and its connection to hope. *Sustainability*, *12*(6), 2181.

Riley, K. and Proctor, L. (2022). A physical education/environmental education nexus: Transdisciplinary approaches to curriculum for a sense of belonging. *Australian Journal of Environmental Education*, *38*(3–4), 267–278. doi:10.1017/aee.2021.29

Riley, K. and White, P. (2019). 'Attuning-with', affect, and assemblages of relations in a trans-disciplinary environmental education. *Australian Journal of Environmental* Education, *35*(3), 262–272. https://doi.org/10.1017/aee.2019.30

Rodler, L. and Renbarger, R. (2023). Strengthening climate change education in the United States. Durham, NC: FHI 360. https://www.fhi360.org/resource/strengthening-climate-change-education-united-states

Samson, F.B., Knopf, F.L. and Ostlie, W.R. (2004). Great Plains ecosystems: past, present, and future. *Wildlife Society Bulletin*, *32*(1), 6–15.

School Strike 4 Climate (2024). Retrieved from https://www.schoolstrike4climate.com/

Sommerfeld, L. (2012). *StressPoints: An overview of water & economic growth in Western Canada*. https://cwf.ca/wp-content/uploads/2015/11/CWF_StressPoints_Report_MAR2012.pdf

Schusler, T.M. and Krasny, M.E. (2008). Youth participation in local environmental action: An avenue for science and civic learning?. pp. 268–284. *In*: *Participation and Learning: Perspectives on Education and the Environment, Health and Sustainability*. Dordrecht: Springer Netherlands.

Stevenson, R.B., Nicholls, J. and Whitehouse, H. (2017). What is climate change education? Curriculum Perspectives, 37, 67–71. doi: 10.1007/s41297-017-0015-9

Tinnell, J.C. (2012). Transversalising the ecological turn: Four components of Félix Guattari's ecosophical perspective. Deleuze Studies, *6*(3), 357–388. doi 10.3366/dls.2012.0070

Guattari's ecosophical perspective. Tytler, R., Bridgstock, R., White, P., Mather, D., McCandless, T., Grant-Iramu, M., Bonson, S., Ramnarine, D. and Penticoss,

A.J. (2019, July 23). 100 Jobs of the future. Ford Australia. Retreived from https://100jobsofthefuture.com/

Verlie, B. and Flynn, A. (2022). School strike for climate: A reckoning for education. *Australian Journal of Environmental Education, 38*(1), 1–12. doi:10.1017/aee.2022.5

White, P.J., Ardoin, N.A., Eames, C., Monroe, M.C. (2023), Agency in the Anthropocene: Supporting document to the PISA 2025 Science Framework, OECD Education Working Papers, No. 297, OECD Publishing, Paris, https://doi.org/10.1787/8d3b6cfa-en.

White, P.J., Ardoin, N.M., Eames, C., Monroe, M.C. (2024). Agency in the Anthropocene: education for planetary health, *The Lancet Planetary Health*, *8*(2), e117–e123, https://doi.org/10.1016/S2542-5196(23)00271-1.

White, P.J., Ferguson, J.P., O'Connor Smith N. and O'Shea Carré, H. (2021). School strikers enacting politics for climate justice: Daring to think differently about education. *Australian Journal of Environmental Education*. 1–14. http://doi.org/10.1017/aee.2021.24

White, P. and Peters, A. (2021). Biology embeds Aboriginal and Torres Strait Islander histories and cultures and applies contemporary Victorian biology research in teaching and learning activities, Science Teachers Association Victoria (STAV)—*Lab Talk 65*(4), 27–34.

Wise, S.B. (2010). Climate change in the classroom: Patterns, motivations, and barriers to instruction among Colorado science teachers. *Journal of Geoscience Education, 58*(5), 297–309. https://files.eric.ed.gov/fulltext/EJ1164623.pdf

Wojcik, D.J., Monroe, M.C., Adams, D.C. and Plate, R.R. (2014). Message in a bottleneck? Attitudes and perceptions of climate change in the Cooperative Extension Service in the Southeastern United States. *Journal of Human Sciences and Extension*, *2*(1), 4. https://scholarsjunction.msstate.edu/cgi/viewcontent.cgi?article=1023&context=jhse

Worth, K. (2021). *Miseducation: How climate change is taught in America*. NY: Columbia Global Reports.

Yazzie, M. and Baldy, C.R. (2018). Introduction: Indigenous peoples and the politics of water. *Decolonization: Indigeneity, Education & Society*, *7*(1), 1–18.

CHAPTER 9

Climate Change

What We Need to Know and How to Inform Others

Kenwyn Cradock

Introduction

The human race is now entering uncharted territory. The global atmospheric carbon dioxide level is currently at 420 parts per million (Ripple et al., 2023). Human actions are causing the planet's climate to change rapidly, i.e., anthropogenic climate change. These changes will have direct consequences for our quality of life and how we survive. To understand what we are facing, we need look no further than the unprecedented climate events of 2023, mainly heat waves (National Aeronautics and Space Administration, 2023) and flooding (CNN, 2023; National Oceanic and Atmospheric Administration, 2023) (Ripple et al., 2023). So, what is climate change? It is a term that everyone has heard, but what does it really mean? One common point of confusion in understanding the issue of climate change is how *climate* differs from *weather*.

While related, the two terms are not synonyms and cannot be used interchangeably. We experience weather whenever we set foot outdoors. Weather varies throughout the day and from day-to-day. Climate,

University of New Mexico, Albuquerque, New Mexico.
Email: kcradock@unm.edu

however, refers to long-term averages (e.g., minimum and maximum tempertures, humidity, rainfall) for an area, typically 30 years or more. Weather data collected over thirty or more years is used to calculate the climatic averages for a region. Thus, climatically, January in the northern hemisphere will be colder than July, while the weather on a particular day may be colder or warmer than the climate average expectations. We do not expect climatic conditions to shift rapidly (i.e., hour-to-hour or day-to-day) (National Oceanic and Atmospheric Administration, 2016). Rather, climate change is the systematic shift in climate parameter averages (temperature, rainfall, humidity, ocean heat, sea level, wind, etc.) over a multi-year period (National Oceanic and Atmospheric Administration, 2016).

Another area of possible confusion is how the terms climate change, and *global warming* are used. Global warming is a component of climate change. While warming of the planet and its atmosphere is indeed happening, climate change also considers several other aspects of the climate (National Oceanic and Atmospheric Administration, 2016). The warming of the planet is not uniform, it occurs more rapidly at the poles than elsewhere. Patterns of precipitation shift in response to more energetic environments (heat is a form of energy). Rain events in some areas may be more intense, delivering more rain in shorter periods of time. Winters may be colder in some areas than they have been historically; summers will on average become warmer.

The basic science underlying why climate change is occurring is well established as it follows the laws of physics and chemistry (National Aeronautics and Space Administration, 2018; Environmental Protection Agency, 2023c). What is far more challenging is trying to understand the impacts of a warming atmosphere on the planet. For example, *How will a warmed atmosphere impact cloud development? How do different types of clouds influence radiation, both incoming and outgoing? What are the tipping points for various biological, geological, meteorological, and climatological systems?* What is known is that once tipping points are crossed, it becomes impossible, within any human-relevant timeframe, to reverse the outcome. There is no *Oops, let's fix that* option.

What Causes Global Climate Change?

Climate change encompasses multiple factors, from local to global. Both human activity and natural phenomena can influence the climate. This is one of the contentious issues in some countries when addressing the climate crisis. While natural events can influence the climate (e.g., volcanic eruptions; La Niña, El Niño and others), such events, over the past few decades, cannot account for the degree of climate shift that has occurred. So, we must consider the role of human actions. Much of the impact stems from the consequences of the industrial revolution and the increased consumption of fossil fuels (coal, oil, natural gas) along with increased urbanization and agricultural activities (National Oceanic and Atmospheric Administration, 2016)

The composition of the atmosphere has a strong influence on the climate of an area. With regards to atmospheric warming, a few gases in particular play a major role. These are primarily carbon dioxide (CO_2), methane (CH_4), nitrous oxide (N_2O) (Ripple et al., 2023) and water vapor (H_2O) - the most abundant of the greenhouse gases (Buis, 2022). Water vapor amplifies the impacts of gases such as CO_2 and CH_4. Warmer air causes more evaporation and holds more water vapor. Water is also less likely to condense and precipitate out of warmer air.

Water has a high *specific heat*. This means that it is good at absorbing and holding heat, which leads to a warmer atmosphere and more evaporation. What we have now is a positive feedback loop. The increased water in the atmosphere influences the global water cycle, making some areas drier and others wetter. The energy contained in water vapor also contributes to more intense precipitation events, contributing to flooding (Buis, 2022). Additional factors include the amount of ice (influences the albedo of the planet), the amount of forested land, and sea temperatures. Another anthropogenic gas we need to consider is sulfur dioxide (SO_2), a compound that results in cooling rather than warming (Ripple et al., 2023). Sulfur dioxide (and related sulfur oxides produced through burning fossil fuels and industrial processes) can influence health as well as the environment. Environmentally, these compounds can limit plant growth, contribute to acid rain (when the gas dissolves in rainwater), and reduce visibility by causing haze. Actions taken by the US Environmental Protection Agency (US EPA) and related agencies around the world have reduced atmospheric concentrations of this gas (Environmental Protection Agency, 2023a).

The changes in the atmospheric concentrations of greenhouse gases have resulted in a marked increase in average global temperature of 1°C (1.8°F), with much of this change occurring after 1970 (Royal Society, 2020). There are some interactions to be aware of. Because of a warming atmosphere there has been a reduction in both, land and sea ice. This reduces the albedo of the planet, meaning that less radiation is reflected back into space and that the darker sea and land absorbs more heat from the sun. Both outcomes result in more heating. Melting land ice contributes to rising sea levels, while melting sea ice decreases the salinity of the ocean. Warm water occupies more space than cold water. So, as the water warms, it contributes to sea level rise (Physics Outreach, n.d.).

Considering anthropogenic climate change, the impacts are fairly straight forward to comprehend. The industrialization of developed economies, something that developing countries are striving to achieve, is powered by the consumption of fossil fuels. They are called fossil fuels because they are made from the decomposing remains of organisms that lived and died in the ancient past, often during the Carboniferous era. The plants that lived during this time utilized carbon dioxide from the atmosphere for photosynthesis, ultimately incorporating much of the atmospheric carbon into their physical structures. When they died, they often fell into swampy environments, as the Earth was then a much hotter and wetter place. Over time, due to the low oxygen environments in these swamps, the plants did not decompose completely but became the oil and coal beds of today (Berner, 2003; Feulner, 2017).

Today we extract oil and coal for the energy reserves stored in these substances, which we harness through combustion. The combustion process then releases carbon dioxide back into the atmosphere. While this is a simplified version of events, it captures the essence of what is happening. We have artificially accelerated the carbon cycle, but not equally across all stages of the cycle.

A natural question at this point is *If plants removed the carbon dioxide all those millions of years ago, why can't/isn't that happening now?* The process of photosynthesis is occurring as it always has. The problem is that we are removing plants at unprecedented rates (Ripple et al., 2023) while also introducing additional carbon dioxide into the atmosphere at a rate that exceeds the capacity of photosynthetic organisms to remove and utilize it.

What is the Evidence?

The evidence that climate change is occurring and that human activity is a major driver is simply overwhelming. Such data is amply covered in publicly accessible reports produced by entities such as the Intergovernmental Panel on Climate Change (IPCC), the National

Oceanic and Atmospheric Agency (NOAA), and the United Nations Environment Programme (UNEP) to name a few.

There are some simple predictions that can be (and have been) made that evaluate the influence of increasing greenhouse gases in the atmosphere and the associated shift in temperature. Hotter days should become more common, along with flooding and extreme rain events. Ice cover across the planet will decrease and sea levels will rise. As humans have been recording weather (and climate) over decades (centuries for some metrics) there are historical records that we can use for comparison. Through comparison, we can see that the mentioned predictions are supported by the data. For a summary of some of these findings, see Figure 3 in Ripple et al., 2023.

Complications for Society

The climate is changing. So what? The predicted outcomes for climate change indicate that humanity will face multiple challenges in the years and decades to come. There will be massive human suffering across the planet as the impacts take place and become the new norm. These hardships will be in the form of increased incidences of disease; reduced productivity leading to reduced food supplies; increased drought and flood events and resulting infrastructure damage, impacts on food production and disease outbreaks; loss of biodiversity; economic hardships related to reduced productivity and increased disease. The specifics and significance of each of these outcomes will be different depending on geographic location, but they will be experienced by all.

Health: Increased heat is itself an issue causing conditions like heat stroke and dehydration that can in turn trigger strokes and heart attacks. Additionally, the changing climate is increasing the development of insects and related organisms, including those that have vector diseases such as mosquitoes and ticks. It is also allowing them to expand their range to areas where they haven't previously been able to live, in turn exposing people and animals to pathogens that they have not historically had to contend with.

Economics: Climate change will impact economics from the local to the global level, but specific impacts will vary by locality and with time scale, i.e., short-term versus long-term costs. We need responsive economic models that can incorporate the complexities of geography and human needs. Such complexity poses many challenges. Sea level rise is going to be a significant factor in coming years (Desmet and Rossi-Hansberg, 2021). The scale of the impacts could be consequential: even with prompt action to meet existing agreements such as the Paris Agreement, we are facing a drop of over 4% in global GDP (Gross Domestic Product); without meaningful action now, the loss could be as great as 18% by mid-century. All countries will be impacted, with developing nations most at risk, while developed nations in the northern hemisphere are more likely be able to respond to the economic challenges (Swiss Re Institute, 2021).

Agriculture: The impact of climate change on agriculture is varied. In some areas conditions will allow for longer growing seasons and the potential to grow crops that previously were not viable in the area. That said, other areas will experience more flooding or drought making it more difficult to produce viable crops. Longer growing seasons and increased temperatures will also increase the challenges posed by weed species and insect pests, while also potentially increasing the need for more irrigation. Changes in how we approach agricultural production can potentially reduce these negative impacts (Environmental Protection Agency, 2023b).

Biodiversity: The rapid shift in climate conditions is creating challenges for biodiversity. In the oceans, increased acidity due to CO_2 dissolving in the water is a threat to coral reefs. These structures are essential for many marine species, including some that are of economic importance. Corals produce exoskeletons consisting of calcium carbonate sensitive to more acidic environments. A more acidic ocean makes it more difficult for the corals to build and maintain their exoskeletons. Additionally, warmer ocean temperatures can lead to coral bleaching. This occurs when the corals eject the symbiotic zooxanthellae that give them their color. Without the zooxanthellae the corals will die (Colbert, 2023; National Ocean Service, n.d.). Warmer waters also increase the risk of disease in corals (National Ocean Service, n.d.).

Terrestrially, warmer temperatures are pressuring organisms to move away from the equator and further up mountain sides as they follow their preferred temperatures. For organisms that cannot move, the increased temperatures can be lethal. The ability of organisms to migrate in accordance with their required environmental conditions may be limited by the height of mountains, and the presence of cities and other developments.

The climate shifts taking place are also influencing phenology; the timing of organisms' annual cycles (Climate Adaptation Science Centers, 2016). While organisms can respond and make changes in their phenology, not all organisms respond in the same way. This can disrupt the interactions among organisms. A further challenge is when some organisms in a system are primarily responding to temperature (which is changing) and others are primarily responding to a parameter such as day length (which is not changing). This can result in a mismatch in the timing of biological events. For example, plants and insects become active earlier in the spring as they respond to temperature, but migratory birds that are responding to day length do not shift their spring migration time and so miss the food resources provided by the plants and insects. This in turn can impact the survival and reproductive success of the birds (Climate Adaptation Science Centers, 2016; Cohen et al., 2018; Ettinger et al., 2022).

Actions We Can Take as Individuals

There is much to be concerned about when it comes to the global climate crisis. It can seem overwhelming. What can an individual contribute to addressing such a monumental challenge? The answer, in fact, is quite a bit! While any one individual changing a habit or two may not achieve much, when many individuals make changes to their lifestyles to reduce their impact on the climate (and environment in general) the results can be significant. Many of these changes are easy to make and may even save money or have other benefits. Below are some ideas about how we can change our routines and habits to reduce our individual contributions to climate change. This information comes from the following sources: United Nations Environment Programme (2022) and the US Environmental Protection Agency (2022). There are, however, numerous resources that can be accessed for ideas and guidance.

1) Change our driving habits to reduce (ideally avoid) unnecessary travel powered by fossil fuels. Use public transportation, walk, or use a bike. Carpool if possible and rather than make multiple trips, try to bundle your activity to achieve multiple ends in a single trip. Avoid using the drive-through as idling consumes fuel while reducing air quality in the immediate vicinity.
2) Reduce our electricity usage. Adjust your housing climate controls (warmer in summer, cooler in winter). Even a few degrees will decrease energy used. Turn off lights and appliances that are

not in use. If possible, unplug devises that are not being used to avoid phantom power draw. Make sure your insulation, windows and doors are in good shape as this will reduce your heating and cooling needs.

3) Switch to a more plant-based diet. This will reduce agricultural contributions to climate change and improve health with regards to conditions such as cancer, heart disease, diabetes, and stroke.
4) Purchase local produce and products when possible. This supports local farmers and businesses while reducing the contributions of transportation to climate change. Purchasing seasonal produce also contributes to these benefits.
5) Reduce food wastage. By managing portion size and not preparing/purchasing more food than we consume, we can reduce waste and maximize the economic benefit from our purchases.
6) Adapt our clothing choices. In consumer-based societies, we have a throw-away culture where we buy clothes for short-term use and then discard them. If we buy fewer clothes of better quality and use them until they wear out, we reduce the impact on the environment. Consider wearing clothes made from renewable/sustainably sourced materials. Using clothing exchanges (or similar entities) allows us to keep a dynamic wardrobe while reducing waste.
7) Reuse or repurpose items. Resealable plastic/glass containers can be used for storage. Plastic shopping bags can be used for storage or as trash bags. Take reuseable bags when you go shopping to reduce the number of disposable shopping bags that you bring home. Buy used and/or donate items when possible.
8) Use water efficiently. Turn off faucets when water flow is no longer needed. Reduce the amount of time you spend in the shower, and/or use low-flow shower heads. If you have a garden, use strategies that are appropriate to your region. Use plants that are native to the area as these will typically require less watering and overall care. They also often have the additional advantage of supporting local wildlife such as pollinators and birds.
9) Continue to educate yourself and others about how to be climate and environmentally responsible. Get engaged with local community groups that are active in these areas.
10) Consider engaging your political representatives from the local to the national level on these topics. Encourage them to consider the climate and environmental consequences of their actions and the legislation that they support.

11) If such changes are new to you and your family, I recommend starting with one or two changes that you think will be easy to make, and then adding more changes later. Trying to take on too much change at once can prove to be challenging and result in feeling discouraged and abandoning the effort. Also, be sensible about how you approach your chosen changes. If you need lights for security then it makes sense to keep those on, rather than risking physical safety by turning off a porch light for example. Introduce one or two plant-based meals per week rather than trying to become vegetarian overnight. As you become comfortable with the changes you are making, expand them (e.g., adding more plant-based meals to your menu) and add new environmentally friendly activities to your repertoire.

Conclusion

It is essential that our global society be aware of and understand the implications and consequences of climate change, which now presents us with a climate crisis. The world that humanity will be living in fifty years from now will be a different place climatologically. How different, depends on our actions today and the technological advancements we can achieve to address climate shifts. The lack of meaningful action by countries across the globe, unfortunately has made the damage far greater than it needed to be, endowing today's children and future generations a world never before experienced by human beings. The exact impacts on the quality (and quantity) of life cannot be predicted in any detail, but we can be reasonably certain that both will be lower for most individuals across the planet.

While there is much room and cause for pessimism, there is also room for optimism. Humanity can be very creative in developing new and wonderful ways for improving life. This is especially true in times of crisis. As it is said, *Necessity is the mother of invention.* We can anticipate many disruptive technologies in the coming decades that will change what humanity understands and can do about the climate-based challenges that we are facing and yet to face. Such optimism is predicated on there being an educated and informed public. This in turn is dependent on robust educational systems where teachers are free to access and utilize the most up-to-date discipline-based information and pedagogical methods to create and develop the minds of coming generations. Climate change is an excellent topic for engaging students with STEM (Science Technology Engineering and Mathematics) fields, along with economics and social studies. There are many opportunities for teaching across the curriculum and showing how different disciplines interact in meaningful and engaging ways. Educational systems at all levels, from early childhood through advanced graduate study, must be free from ideological interference if we are to have a totally free society that has the intellectual tools necessary to accept and conquer the challenges posed by the climate crisis that is upon us.

References

Berner, (2003). *The long-term carbon cycle, fossil fuels and atmospheric composition.* Nature 426, 323-326

Buis, A. (2022). *Steamy Relationships: How Atmospheric Water Vapor Amplifies Earth's Greenhouse Effect.* National Aeronautics and Space Administration. https://climate.nasa.gov/explore/ask-nasa-climate/3143/steamy-relationships-how-atmospheric-water-vapor-amplifies-earths-greenhouse-effect/).

Climate Adaptation Science Centers (2016, December 31). *The Impacts of Climate Change on Phenology: A Synthesis and Path Forward for Adaptive Management in the Pacific Northwest.* https://www.usgs.gov/programs/climate-adaptation-science-centers/science/impacts-climate-change-phenology-a-synthesis

CNN (2023, September 17). *Ten countries and territories saw severe flooding in just 12 days. Is this the future of climate change?* https://www.cnn.com/2023/09/16/world/global-rain-flooding-climate-crisis-intl-hnk/index.html

Cohen, J.M., Lajeunesse, M.J., Rohr, J.R. (2018). *A global synthesis of animal phenological responses to climate change.* Nature Climate Change 8, 224–228. https://doi.org/10.1038/s41558-018-0067-3

Colbert, A. (2023). *Vanishing Corals, Part two: Climate Change is Stressing Corals, But There's Hope.* https://climate.nasa.gov/explore/ask-nasa-climate/3290/vanishing-corals-part-two-climate-change-is-stressing-corals-but-theres-hope/

Desmet, K., Rossi-Hansberg, E. (2021). *The Economic Impact of Climate Change over Time and Space.* The Reporter, No. 4, December. https://www.nber.org/reporter/2021number4/economic-impact-climate-change-over-time-and-space

Environmental Protection Agency (2023a, February 16). *Sulfur dioxide (SO_2) pollution.* https://www.epa.gov/so2-pollution/sulfur-dioxide-basics

Environmental Protection Agency. (2022, December 27). *What You Can Do About Climate Change.* https://www.epa.gov/climate-change/what-you-can-do-about-climate-change

Environmental Protection Agency. (2023b, November 16). *Climate Change Impacts on Agriculture and Food Supply.* https://www.epa.gov/climateimpacts/climate-change-impacts-agriculture-and-food-supply

Environmental Protection Agency. (2023c, November 01). *Basics of Climate Change.* https://www.epa.gov/climatechange-science/basics-climate-change

Ettinger, A.K., Chamberlain, C.J. and Wolkovich, E.M. (2022) The increasing relevance of phenology to conservation. Nature Climate Change 12, 305–307. https://doi.org/10.1038/s41558-022-01330-8

Feulner, G. (2017). *Formation of Most of our Coal Brought Earth Close to Global Glaciation.* Nature *114*(43), 11333–11337. https://doi.org/10.1073/pnas.1712062114

National Aeronautics and Space Administration. (2018, June 06). *The scientific method and climate change: How scientists know.* https://climate.nasa.gov/news/2743/the-scientific-method-and-climate-change-how-scientists-know/

National Aeronautics and Space Administration (2023, September 14). *NASA Announces Summer 2023 Hottest on Record.* https://climate.nasa.gov/news/3282/nasa-announces-summer-2023-hottest-on-record/ (accessed April 03, 2024)

National Ocean Service. (n.d.). (part of the National Oceanic and Atmospheric Administration) *Corals Tutorial.* https://oceanservice.noaa.gov/education/tutorial_corals/welcome.html (accessed November 26, 2023).

National Oceanic and Atmospheric Administration. (2016, March 09). *What's the difference between climate and weather?* https://www.noaa.gov/explainers/what-s-difference-between-climate-and-weather

National Oceanic and Atmospheic Administration (2023, August 8). *July 2023 brought record-high temperatures, devastating floods across the U.S.* https://www.noaa.gov/stories/july-2023-brought-record-high-temperatures-devastating-floods-across-us (Accessed April 03, 2024).

Physics Outreach (n.d.). *Albedo and Climate Change.* University of Wisconsin. https://wisc.pb.unizin.org/climatechange/chapter/chapter-1/ Accessed on November 27, 2023.

Ripple, W.J., Wolf, C., Gregg, J.W., Rockström, J., Newsome, T.M., Law, B.E., Marques, L., Lenton, T.M., Xu, C., Huq, S., Simons, L., King, D.A. (2023). *The 2023 state of the climate report: Entering uncharted territory.* BioScience https://doi.org/10.1093/biosci/biad080

Royal Society (2020). *Climate Change: Evidence and Causes, Update 2020.* The Royal Society and US Academy of Sciences. https://royalsociety.org/~/media/Royal_Society_Content/policy/projects/climate-evidence-causes/climate-change-evidence-causes.pdf

Swiss Re Institute. (2021). *The Economics of Climate Change: No Action Not an Option.* https://www.swissre.com/dam/jcr:e73ee7c3-7f83-4c17-a2b8-8ef23a8d3312/swiss-re-institute-expertise-publication-economics-of-climate-change.pdf

United Nations Environment Programme (2022, May 04). *10 ways you can help fight the climate crisis.* https://www.unep.org/news-and-stories/story/10-ways-you-can-help-fight-climate-crisis

Index

Printed in the United States
by Baker & Taylor Publisher Services